# COUNTDOWN

# COUNTDOWN

## The Blinding Future of Nuclear Weapons

## SARAH SCOLES

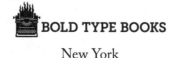

BOLD TYPE BOOKS

New York

Bold Type Books
Hachette Book Group
1290 Avenue of the Americas, New York, NY 10104
www.boldtypebooks.org
@BoldTypeBooks

Printed in the United States of America

First Edition: February 2024

Published by Bold Type Books, an imprint of Hachette Book Group, Inc. Bold Type Books is a co-publishing venture of the Type Media Center and Perseus Books.

The Hachette Speakers Bureau provides a wide range of authors for speaking events. To find out more, go to hachettespeakersbureau.com or email HachetteSpeakers@hbgusa.com.

Bold Type books may be purchased in bulk for business, educational, or promotional use. For more information, please contact your local bookseller or the Hachette Book Group Special Markets Department at special.markets@hbgusa.com.

The publisher is not responsible for websites (or their content) that are not owned by the publisher.

Print book interior design by Amy Quinn.

Library of Congress Cataloging-in-Publication Data

Names: Scoles, Sarah, author.
Title: Countdown : the blinding future of nuclear weapons / Sarah Scoles.
Other titles: Blinding future of nuclear weapons
Description: First hardcover edition. | New York : Bold Type Books, 2024. | Includes bibliographical references and index. |
Identifiers: LCCN 2023028216 | ISBN 9781645030058 (hardcover) | ISBN 9781645030072 (ebook)
Subjects: LCSH: Nuclear weapons—Government policy—United States. | Nuclear weapons—United States—Forecasting. | Nuclear weapons—United States—History. | Nuclear weapons—United States—Safety measures.
Classification: LCC UA23 .S377 2024 | DDC 355.8/251190973—dc23/eng/20230824
LC record available at https://lccn.loc.gov/2023028216

ISBNs: 9781645030058 (hardcover), 9781645030072 (ebook)

LSC-C

Printing 1, 2023

# CONTENTS

# INTRODUCTION

I'VE NEVER BEEN PART OF AN AIR-RAID DRILL. *THE DAY AFTER* WASN'T my primetime special. I only knew the USSR was a thing because my Florida elementary school didn't have the budget for new maps for a few years after the Soviet Union's collapse. I had nightmares about falling into sinkholes but not about nuclear-tipped missiles streaking through the sky.

In fact, I didn't think about nuclear weapons much until, many years after those nightmares stopped, I moved out West. Here, in the Rocky Mountain region, the bombs—or at least their legacy—were a little harder to ignore. I initially lived a short drive from Rocky Flats, a plant that once produced the cores of nuclear weapons. Down the interstate from there is Cheyenne Mountain, a nuclear command center built beneath thousands of feet of granite, meant to withstand thermonuclear blasts so the people inside could launch missiles even if the rest of us died. This part of the country is scarred by old uranium mines and mills. Missile silos and alert centers, current and former, carve out the earth.

This region is also home to major nuclear research and production facilities, specifically Sandia and Los Alamos national laboratories. Weapons labs. When I moved here, I knew of the facilities' existence but not much about what happened behind their extremely closed gates. Being a journalist who writes about science, though, I was able to simply ask if I could drop by.

On my first trips to both labs, I spoke to scientists who constructed particle accelerators, simulated the start of the universe, made next-generation supercomputers run, and replicated supernovae, among other far-out topics. It was all interesting, sure—but I knew none of this research was the *reason* these labs existed or even the full story about these particular scientists' specialties. The national labs run by the Department of Energy's National Nuclear Security Administration (NNSA)—Sandia and Los Alamos, as well as Lawrence Livermore in California—like to discuss the easily palatable, press-releasable science their employees do.

For one, it's not classified. For two, it doesn't make people mad.

But the scientific forces animating their research also show up in nuclear weapons. They are the labs' raison d'être, and the majority of their budgets go to *that* sort of stuff, which they call "mission work": research on and maintenance of nuclear weapons and national security.

Yet that work is kept more under wraps, a tendency that has kind of matched public awareness of nuclear weapons, which receded after the Cold War. But the United States is currently in the middle of a giant nuclear modernization program, reinvesting in its atomic infrastructure like it hasn't in decades. And as tensions have risen globally, particularly with Russia's invasion of Ukraine, nuclear weapons' salience to the modern world is seemingly here to stay. As such, the quiet parts deserve to be discussed out loud. *Countdown* is an attempt to turn up the volume on those quiet parts, as they play out among the scientists at nuclear labs, and illuminate the place nuclear weapons hold—in the world, in the country, in scientific research, for the people who work on them, and for those of us who don't.

# Part One

# CRITICAL MASS

AMERICA'S NUCLEAR ARSENAL IS OLD: THAT'S A FACT. AMERICA'S NU-
clear arsenal is *too* old and needs significant modernization: about that,
reasonable people can, and definitely do, disagree.

Despite those disagreements, nuclear bombs are nevertheless getting
age-altering surgery—nips and tucks and new parts. A lot of new parts.

New supercomputers and advanced simulations are, at the same
time, trying to understand the inner workings of the weapons. New
sensor-fusing systems are improving and quickening nuclear explosion
detection.

With all this, the National Nuclear Security Administration labs are
in a bit of flux, their role more active and more public than it has been in
recent years. As they move into a new nuclear age, along with the rest of
us, history presses into the present and points toward an uncertain future.

# CHAPTER ONE

We knew the world would not be the same.

> —J. Robert Oppenheimer, on the
> test of the first atomic bomb

THE SLOPES OF PAJARITO MOUNTAIN SKI AREA sit two thousand feet and about fifteen minutes above the New Mexico town of Los Alamos. If you looked down from the lift, the city would seem small. Inconsequential, even. The main offices of Los Alamos National Laboratory would look even smaller, like someone threw Legos on the ground and lit them with high-powered lamps.

But it's all about perspective: in reality, Los Alamos National Laboratory—called LANL, pronounced like *flannel* minus the *f*—is a more than $4.6-billion-per-year operation that occupies almost forty square miles of high-altitude plateau. The lab's stated mission is "to solve national security challenges through simultaneous excellence."

It's unclear, linguistically, what "simultaneous" means, but it's probably classified anyway. LANL, after all, works on some of the most tightly held US secrets: those surrounding nuclear weapons.

The city of Los Alamos was founded in the early days of the Manhattan Project, the US initiative to build the world's first atomic bomb. Back then, the town's existence was hushed. Home to the code-named "Project Y," and so sometimes called Site Y along with Atomic City, Magic Mountain, Shangri-La, and probably a few French words, it was hidden not just from the American people but also from Congress. Its physicists toiled in obscurity to figure out how to harness the sort of power that nature had managed to master in a much earlier cosmic era. Those physicists—living in hastily constructed homes, with few amenities and little privacy from each other, even if they had almost infinite privacy from everyone else—eventually succeeded in explosively splitting atoms. And in so doing, they changed life not just on the Pajarito Plateau but also on the whole planet.

From a perch on the mountain above town, it's easy to map the landforms bordering it. Los Alamos sits on four mesas, so cliffs prevent it from exceeding its present stretch. LANL may be able to change the world, but it has not gone so far as to alter this geophysical fact.

Looking through the relatively apolitical calm of aspens that try their best to block the view, you have to remind yourself of what this place is. Of how a settlement that was, in its heyday, much smaller and more secluded than it is today nevertheless shifted Earth's trajectory. You have to reach to remember that ghosts working with kludgy code names and poor municipal infrastructure helped bring into being a kind of weapon that could destroy everyone on the planet, with relative ease. The push of some buttons. A few correct codes. A presidential whim. A moment of panic. A straight-up mistake. A misperception. "If you think about how many of these actual nuclear weapons are out there, it's insane that we haven't had an accident or haven't gotten into a hot war," says Aaron Arnold, a former fellow with the Project on Managing the Atom.

But if, for whatever reason, a bomb does detonate in a conflict once again, there will be no returning to the world as it was before. "Once a country starts firing them off, that's kind of it," says Arnold. The clock can't run backward and undo the first strike, and so another country, or a plurality of other countries, will likely respond with atomic retaliation.

"In all likelihood, that's it," Arnold says again, and pauses. "That's it."

Regardless of whether the "it" eventuates, this planet will always bear the imprint of the deadly device first devised in the flat spaces between

Los Alamos's gaps. Those scientists didn't, of course, do it alone. And if they hadn't done it, someone else would have, at some point.

But they did do it, and they did it here, and here we are.

LANL, in the twenty-first century, largely exists to manage what it and other Manhattan Project facilities wrought decades ago. Run by the Department of Energy's National Nuclear Security Administration (NNSA), the lab is tasked in part with helping maintain and modernize the United States' stockpile of nuclear weapons—ensuring, as employees are fond of saying, that the bombs always explode when they're supposed to, as they're supposed to, and never explode (or do anything else) when they're not supposed to. This idea gets shortened to the quip "always, never," to the point where people just say, "You know, always, never." They also know that someone who says the phrase "safe, secure, and reliable" is talking about the desired traits not of a car or a romantic interest but of thermonuclear arms.

While LANL is no longer the city's only employer, it remains the biggest, and the community maintains the feel of a company town. Perhaps that has something to do with the county logo, where the canonical atomic symbol sits inside *Los*'s *o*. The electrons' elliptical orbits focus the eye on where the nucleus, the center of power within an atom—and here in town—would be.

Or maybe it has something to do with the old guard-shack replica—now a public bathroom—you pass as you enter the city limits. The original was just a small, white structure where people presented passes before proceeding. The identical building on which the replica is modeled kept Los Alamos from the public during the Manhattan Project. So, too, did the muddy, rutted road leading up from Santa Fe, which few would have found themselves traversing by accident or for fun.

Today, obviously, Los Alamos isn't a secret. A many-lane blacktop now leads directly to Site Y. Commuters can scream up and down the hill easily, and so can information: the lab puts out press releases about its more innocuous work, its research on everything from pandemics to planetary science. The town has a McDonald's. There's a Starbucks, which expanded to a new, stand-alone building in 2022. You can even pick from a couple of craft breweries. A coworking space cheekily sports the original code name for this part of the Manhattan Project. There's even a national historic park, appropriately called Manhattan Project National Historical Park, dedicated to remembering how the bomb was built here.

On Pajarito hill, the Manhattan-era leader of the original project, J. Robert Oppenheimer, used to enjoy a relative lack of friction and a good view when he ascended to that vantage point to ski. That view is quite different today, but the setting sun still sets the Sangre de Cristo Mountains on fire, just as it did in Oppenheimer's 1940s, fusing hydrogen the whole way down to the horizon.

As the star descends, empty areas between the mesas go dark; only Site Y's interstitial spaces beam into the evening sky, showing a proverbial city on a hill or a scourge, depending on whom you ask.

Regardless, it's a place whose emissions reveal its existence, no matter what.

The result of the Manhattan Project and Los Alamos's early research was a bomb the scientists dubbed the Gadget. No one knew for sure whether—or exactly how—it would work. We still don't understand exactly how it or any of the thousands of other bombs we've since made function. The only thing that mattered, back then, was that they worked.

The day of its test, an event dubbed Trinity, many of the lab's scientists trekked out to the site, near the New Mexico town of Alamogordo, to watch their creation (hopefully) destroy itself. Colorful tales of the day abound: Physicist Edward Teller passed around suntan lotion, anticipating the radiation they'd receive like a day at the beach. Enrico Fermi got ready to drop scraps of paper to estimate the strength of the blast wave, as they flew away on the wind. A broadcast of the countdown crossed frequencies with a local radio station, and the *Nutcracker* suite superimposed itself over the Trinity test's "three . . . two . . . one."

When the clock reached zero, the orchestra played on. But at the Trinity test site, a fireball erupted, blinding the researchers in the first microseconds. The closest observers were just six miles away. A column of smoke soon rose into a roiling mushroom cloud. The cloud, levitating higher and higher, seemed to suck up the earth itself, siphoning resources as it ascended toward heaven. It was purple, blue, red, violet. Violent. Glowing. Thundering. Awesome. Terrifying.

People saw the effects of the blast in three states. Those living close by wondered whether this was some kind of apocalypse. The army informed them—lying—that it was just an accidental explosion at a munitions storage area.

But the residents weren't wrong in their assessment. As chemist George Kistiakowsky said after witnessing Trinity, "I am sure that at the end of the world—in the last millisecond of the earth's existence—the last man will see what we saw!"

The subatomic particles ruling the bomb follow the laws of quantum mechanics, which in popular, if slightly incorrect, understanding means that they are uncertain: in multiple places at once, doing multiple things, only settling into one state when you look at them directly—like Schrödinger's cat, a being both alive and dead.

The scientists, too, occupied several states simultaneously that day, a mix of feelings sluicing through their bodies when the Gadget erupted as intended: Pride. Fear. Regret. Joy. Anticipation.

Most of all, power. They had taken a theoretical, abstract science and alchemized it into something devastatingly concrete. In the minds of some of the researchers, this ultimate weapon would bring with it ultimate peace, because it was too horrible to ever use (after, of course, their country used it twice), especially if you knew a similar weapon could be deployed against you in response. That's part of the basic ideology of deterrence: we have the weapons so that no one will horribly attack us, or our allies, because they know we could attack horribly back. And other countries' weapons keep us from the temptation to horribly attack them. Nuclear weapons, then, keep large-scale wars from breaking out.

Similar deterrent ideals echo like a hymnal chorus, in eerily similar language, from scientists and engineers across LANL and the other weapons labs, Sandia and Lawrence Livermore national laboratories. Workers at these sites sometimes clutch the words like verbal talismans: "always, never," "a credible deterrent." But deterrence, like any other philosophy, is a belief, and one that is impossible to test, because you cannot wind back the doomsday clock and see how various conflicts and dynamics would have played out if fission and fusion had remained scientifically opaque.

For the record, the inventor of dynamite also thought its existence would halt huge conflict, for similar reasons. And yet here these Manhattan scientists were, decades later, devising something much worse, repeating history as humans seem wired to do.

Those scientists have said a lot of quotable things about Trinity in the decades since the test. But in the actual aftermath of the detonation, most, according to Oppenheimer, were silent. Oppenheimer himself thought, famously, of the line "Now I am become Death, the destroyer of worlds," a quote from Hindu scripture.

Less poetic and more to the point was Kenneth Bainbridge, who intoned, "Now we are all sons of bitches."

For almost two decades following Trinity, the spot was marked by a wooden sign labeled "Zero." As in Ground Zero. As if time and space started here.

If you watch nuclear explosions in archival videos today, including footage from Trinity, it's easy to understand how those mid-century scientists could feel their work was that cosmically significant and how they could be both completely undone by seeing such an explosion and obsessed with the idea of it. How they could love it while hating it. These weapons are the raw power of the universe—harnessed, targeted, let loose. They are also beautiful. Staring as the seconds pass, you may feel like you do when standing on the edge of a cliff: A still, small impulse urges you to jump. You're not depressed, and you're not actually going to jump, just like you don't wish for nuclear war. You're simply drawn to the abyss, precisely because of its abyssal nature. That's the pull of a nuclear detonation.

Three weeks after the Trinity test, the United States dropped a replica of the Gadget on the Japanese city of Nagasaki, a few days after dropping a different type of nuclear weapon on Hiroshima. The number of casualties is still unknown, but the high-end estimate holds that around 210,000 people died as a result of both explosions. Some were simply vaporized.

Of the bombing of Hiroshima, the White House's press release said, "We have now won the battle of the laboratories as we have won the other battles."

ALMOST EIGHTY YEARS LATER, NNSA's NATIONAL LABORATORIES ARE still engaged in a forever-war version of that same battle: the fight for nuclear supremacy. Yesterday's United States wanted to have the first nuclear weapon, and today's desires the best radioactive arsenal on earth. Then, the thinking goes, the country can most effectively scare other states with their own arsenals—and those without—away from conflict. The scaring doesn't work, though, unless the other side believes you'd actually pull the trigger. Which functionally means you must be willing to do just that—to "win" if you must, even though this is contrary to Mikhail Gorbachev and Ronald Reagan's famous joint statement: "A nuclear war cannot be won and must never be fought."

Nuclear war has crept closer to the front of the public mind lately. But the American arsenal has never left the minds of many officials, who worry about the expansion and modernization of China's and Russia's weapons, about Russia's aggression and unpredictability, about Iran and its alliances, about North Korea's roguishness, about terrorists with dirty bombs, and about the geriatric nature of this country's own radioactive arsenal. The cost of giving the US nuclear complex a makeover would be around $1.7 trillion over thirty years, according to a 2017 projection from the Congressional Budget Office. There will be a newly designed weapon and alterations and upgraded components for old ones. The Cold War objects will be shined up and readied for a journey through the twenty-first century. Los Alamos and the Savannah River Site, an NNSA facility in South Carolina, will soon restart production of plutonium "pits"—hollow spheres of radioactive metal that form the heart of nuclear weapons. The country hasn't made new plutonium pits on a large scale since the late 1980s.

Some say those developments are necessary, even actively peace promoting, because they uphold deterrence: the weapons as they exist now—aged—aren't as credible or reliable as they could be, so updating and upgrading them may forestall war by more effectively keeping others at bay. Some, meanwhile, say the modernization program and pit production are hawkish and could foment a never-ending arms race.

The truth is probably a bit of both, like Schrödinger's cat.

ARGUABLY THE MOST IMPORTANT PART OF NUCLEAR WEAPONS, PLUTO-nium pits are hollow, eerily silver spheres made of the actinide element 94. They look like some indecipherable artifact from an ancient alien civilization—the kind that tries, and fails, to communicate a message to humanity in an action movie. Their name calls to mind the hard middle of a fruit or a dark hole.

In a bomb, a conventional explosive surrounds the pit. That combustible material goes boom, squeezing the radioactive material together. When the atoms get tight enough, they start to split, initiating fission. With the right conditions, the nuclear reaction sustains itself. And when it gets amped up enough—as it does in modern thermonuclear weapons—that triggers more fission and a lot of fusion, or the combining of atoms, in a secondary part of the bomb. There are many more details, but that's the gist of their inner workings.

The last place the United States made plutonium pits at any sort of scale was the Rocky Flats production plant in Colorado. Situated between Denver and Boulder, today it's known as the Rocky Flats National Wildlife Refuge, and it boasts a wide-open view of both the pancake plains and the saw-toothed Rockies. If you want, you can hike around ten miles of trails on its 5,237 acres. You can also ride a horse or a bike. And in your meanderings, you'll almost certainly see wildlife: mule deer, prairie dogs, jackrabbits, porcupines, hawks, elk, suburbanites in outfits entirely produced by Patagonia. But it took a lot of rehabilitation to bring that friendly outdoor space into being, because the production plant left a serious mark on the land while making those Cold War pits.

During the Cold War, the Rocky Flats Plant, operated by a contractor called Rockwell, whipped up between one and two thousand pits per year. It was thusly productive, until June 6, 1989. On that fateful day, plainclothes FBI agents, whose agency helps enforce some environmental laws, showed up and claimed they wanted to talk to the leadership about an ecoterrorist threat—a believable scenario, given the widespread protesting and activist trespassing by a population of activists who included the likes of Daniel Ellsberg, the Pentagon Papers leaker. But no threat actually existed that day: the agents were simply stalling so that dozens of vehicles and more than seventy armed agents could get to Rocky Flats. As soon as they arrived, the FBI agents told the facility officials why they were really there: to investigate rampant environmental violations. Rockwell later pleaded guilty to charges stemming from violations of the federal Clean Water Act and the Resources Conservation and Recovery Act.

Gotcha.

Soon enough, the feds shut Rocky Flats down and transformed it into a Superfund site. Ten years and $7 billion later, the environmental restoration was finished. And in 2018, the land became the somewhat controversial haven for hikers and hawks that it is today.

Since the site's shutdown, the United States has been missing what Rocky Flats once provided: no other place has produced plutonium pits at scale. The country has relied on existing pits, which have been steadily aging for decades.

That missing capability is why Congress has required NNSA to be able to whip up eighty plutonium pits per year by 2030: to replace the geriatric pits and to create new ones more suitable for modern warheads. "An eighty-pit-per-year capability is a modest and prudent approach to sustaining something as important as maintaining confidence in the US nuclear deterrent," says NNSA's Michael Thompson, a principal assistant deputy administrator.

The pit-production gig is being split between Los Alamos, tasked with making thirty pits per year, and the Savannah River Site, which will shoulder the other fifty. This effort will cost billions of dollars. And it is currently behind schedule by around five years on Savannah River's part. LANL, meanwhile, fell about a year behind during the Covid-19 pandemic.

The plutonium for the project will come from decommissioned weapons whose impotent cores together weigh tens of tons. Here, swords turn not into plowshares but into better swords. The role those swords play, however, and how effectively, is shifting.

The nuclear threats of the twenty-first century are different from those of previous eras, with more competition between more big peer countries, more concerns about smaller countries, and attention to new varieties of radiological terrorism, among other fractious developments. We also have many ways to watch what everyone else is up to: the world is more faceted and more knowable. The war game has changed.

These modern dangers are not lesser than those of the air-raid-drill era: in fact, many experts believe the risk of an international nuclear catastrophe is as high as it has ever been, even during the chilliest temperatures of the Cold War.

If nuclear danger is so clear and present, we could all better understand the culture, science, politics, and people of the current nuclear era—not just those of Dr. Strangelove's.

# CHAPTER TWO

As we headed west on our way toward the hills, we'd pass lines of protesters standing with signs at the gate of the Rocky Flats plant. "Hippies and housewives," my father would grumble, and my mother nodded silently. "Students, too," he'd add. "Wait 'til they get out into the real world."

—Kristen Iversen,
"The Accidental Activist"

Los Alamos National Laboratory (LANL) physicist Tess Light grew up near Rocky Flats back when it was an active production plant. To those nearby, like her, its radioactive mission wasn't a secret: the kids at Light's liberal, experimental school—the Tanglewood Open Living School—were well aware of the plant's activities. They were also unhappy with them. "We didn't have a sports team, but we did protest Rocky Flats," Light says. It is a joke, and it is also true.

Living where they did, in the 1980s, Light and her classmates sensed the shadow of potential apocalypse not just when they watched the evening news with their parents but also when they went outside to play. "I felt," she recalls, "like I was at Ground Zero."

That was a different era, one when nuclear war was a topic for prime-time TV, not one left out of the curriculum till high school, as was sometimes the case in the years after the Soviet Union collapsed. "Me and all my friends from hippie middle school watched *The Day After* and were horrified," Light says. In that network-debuted movie, nuclear war escalates and plays out, with a zoomed-in lens on a small town in Kansas. There, residents deal with and die from the conflict. *WarGames*, too, was popular. In that film, a young hacker nearly starts World War III after accidentally worming into a military supercomputer.

Fictional film scenarios running through their developing minds, Light and her best friend decided to make their own bug-out plan. Their goal in the event of impending nuclear catastrophe was to find each other and then find their moms and ride out the end of the world together. It wasn't detailed—more a statement of priorities than a set of logistics—but the exercise in control calmed them.

Given her hippie school and her fear of nuclear war, it seems a little strange for Light to have ended up at a place like LANL. "My friends would be horrified," she says. The childhood protesters, she means. The idealists. But the real world, as she learned in the years between middle school and the present, is more nuanced than they knew back then.

More nuanced, she thinks, than her old friends might know even now.

"I don't think they want to know what I do," she says. They think, she imagines, "I am in some way contributing to the war machine."

She sees it differently. "I agree wholeheartedly we should not have nuclear weapons," she says. But creating that world is not as simple as taking all the bombs apart, waving white flags, and easing cognitive dissonance and the danger simultaneously. "No nuclear weapons," Light muses, "that's not gonna happen."

Not, she means, any time soon. Not in her lifetime.

"I was born into this world—my children were born into this world—with this horrible threat," says Light. There is no changing the fact that decades ago, physicists at her very lab reproduced the power inherent in an almost insignificant mass of atoms.

The realism tinging Light's statements seems to run through the scientists who end up working at the National Nuclear Security Administration labs: nuclear weapons, whether they help stabilize the world or not, exist, and likely will continue to. Someone has to work on them, and don't you want level-headed people—not war hawks—to do that and to help stop them from proliferating?

At the lab, sure, a few people are creepily fans of nuclear weapons: they'd kind of like to jump off the aforementioned cliff. Then, of course, there are people who believe, in classic deterrence style, that the existence of nuclear weapons stabilizes the world. "I consider that debatable, because I don't see that the world is so stable, right?" Light says. "We found other ways to kill each other. And we just now do it while we have a big threat in the background."

And then there are people who think the best path to stability is less radioactive activity. That's Light's camp. "I'm not sure that humans will ever achieve stability, but nukes are like handing a toddler a lighter," she says. "The kid is going to hurt himself and tear up the house regardless, but without the lighter, you might avoid the kid burning the entire house down before he matures." And if her old friends scrutinized the current principles guiding the Tanglewood School, now called the Jefferson County Open School, they might find she's done a pretty good job of embodying them. The institution's current goals say students should learn to "rediscover the joy of learning, seek meaning in life, adapt to the world as it is, prepare for the world that might be, and create the world as it ought to be."

For Light, that entails some duality: working for the organization that maintains and betters the bombs that she agrees wholeheartedly should not exist.

Light, then, represents a paradox of the modern nuclear weapons complex: now that the bombs exist because of labs like LANL, labs like LANL employ some people who detest that explosive existence and spend their days trying to ensure no more ever go off and that they eventually go away. Even nuclear workers who are in favor of the weapons also believe their existence can lead to peace. And so it is that in each nuclear weapons worker's personal narrative, they're doing the right—safe, secure, reliable—thing. The different, opposing states feel true, simultaneously.

"I hope it matters," says Light, shrugging toward the atmosphere. "That's all I can say."

LIGHT'S BUILDING, DEEP INSIDE THE LANL CAMPUS, BOASTS A BIG Q clearance sign on the front door. There, on a warm day in August 2021, Light blocks the threshold. "I need to verify your citizenship," she says. She'd already sent around an email to everyone in the building, informing them that an uncleared visitor would be around.

A printout on the wall of Light's office shows a plot of data hard to discern from afar. Its lines take the shape of gapped molars. Nearby, a darkened computer screen is plastered with red and white "secret restricted data" stickers. But perhaps most illuminating is a small banner tacked near her workspace that reads, "Je suis une femme formidable."

It's not wrong. Light is a bewitching presence, even with the bottom half of her face hidden behind a mask. She sports flared sleeves and a necklace with a bold, curved pendant. She also writes plays (and occasional letters to the editor of the *Los Alamos Daily Post*) in her spare time. The scripts "tend to incorporate any or all of the following," according to the New Play Exchange: "sarcasm, death, sarcastic death, Buddhism, foodism, poetry, song, and Shakespeare." In one, about lesbian wives who—because quantum mechanics—can see the deceased residents of their house, the staging notes demand that an easel display math. Light is happy to supply examples for interested directors but suggests a quadratic equation with two solutions, a light cone diagram, a spacetime cube, or a partially solved Fourier transform.

Easy.

The Buddhist themes and gallows humor that show up in Light's plays probably help her keep an even mental keel. After all, she deals with the specter of radioactive doom, day in and day out, in her job as project scientist for a sensor that helps detect nuclear detonations from the vantage of space. The goal of such programs is to keep watch from above, using satellites to see if anyone, at any time, is exploding a nuclear weapon aboveground. On any given day, no one *expects* anyone else to be letting loose an incendiary and critical chain reaction (though it is, of course, good to check). The issue is that you need to be able to tell both whether countries are detonating weapons in war and adhering to treaties that ban such tests, like the aptly named Comprehensive Nuclear-Test-Ban Treaty: if you can't decipher whether anyone is breaking such rules, the rules don't really matter. Think of it like speed limit signs on a deserted road one hundred miles from the nearest police officer. To make the treaties themselves credible, and to also provide warning about what other countries might be up to, you have to—to continue the analogy—station some squad cars along that lonesome stretch of highway. Satellite-based systems are some of those squad cars.

A plethora of sensors exist on the ground to manage the prophylactic patrol and catch underground tests. They feel seismic activity, listen for infrasound waves, watch for radioactive particles, and pick up radio waves. The international Comprehensive Nuclear-Test-Ban Treaty Organization maintains such a surveillance network, and the US Air Force hosts a separate one for national use.

Light, meanwhile, helps with the data gathering that LANL does from orbit. Aboard every GPS satellite sit sensors not to tell you how to get to McDonald's but to tell people like Light whether something resembling a nuclear explosion has happened. These global burst detectors are a kind of commensal watchman, along for the ride with the GPS spacecraft's main body, and can pick up visible-light and X-ray emissions, along with the electromagnetic pulses accompanying nuclear detonations that preppers fear. A new experimental platform called STP-Sat-6 also launched in 2021, hosting the Space and Atmospheric Burst Reporting System and an experimental test-detection payload called SENSER. Technology aboard this smaller satellite could form the backbone of future detection systems.

To understand the signals present and future instruments pick up, Light and other scientists must also understand the atmosphere through which they move and how it affects them. For more than a decade, Light and dozens of other scientists have been working on a simulation of the physics of how all different kinds of signals are affected on the journey from their source to the sensor. "We are inches from getting the whole thing signed off on," she says. "It's just so big."

Peering through glasses across the table, only her eyes detectably smiling, she looks on the suspicious side of curious, on the ajar side of open.

LIGHT'S PATH TOWARD THIS RESEARCH BEGAN WHEN SHE WAS EIGHT, IN the car with her father and siblings. "He's like, 'So kids, what do you think you want to do when you grow up?'" she recalls. "And on the spot, I just was like, 'I don't know, I want to be an astronaut.'"

She'd never thought particularly much of space, but like an entrée choice you didn't know you were considering at a restaurant, the idea simply dropped out of her mouth, unexpectedly. It continued to guide her trajectory, which is how she ended up in a high school physics class.

She hated it—but continued. "What a shame that I'm going to do this for the rest of my life," she thought. "It never occurred to me to do something that I like."

And so she continued into college, studying physics so she could become an astronaut. But this time, taking courses on topics like elementary particle physics, she started to love it. And she wanted to apply her physics skills to astronomy. "It's just kind of like this weird Brownian walk to get somewhere," she says, referring to a phenomenon where particles suspended in a medium move randomly about.

In graduate school, though, trouble clouded the view again. "You have to compete for time on telescopes, and you have to write about why it's really crucial to the field that they give you the time on the telescope," she says. When she reached for her reasons—like that someone needed to observe a given galaxy in the next million years or they would not be able to view it anymore—she found them lacking. "I just couldn't muster the sense of urgency," she says.

It's a common refrain from people who end up working at the three weapons labs, which insiders sometimes call the "tri-labs." These facilities are focused on science in the service of national security, nuclear or otherwise. And their scientists want to do something that matters more immediately to the people of Earth than adding to basic human knowledge—noble and necessary as that may be. "The applications give me a sense of purpose, while pure, esoteric knowledge feels cold," explains

Light. "But that's just me, and thank god many people feel differently or we'd never advance at all."

Around the time Light was finishing school and fishing for a job, a friend who'd worked for Los Alamos suggested she look at the lab. Light began exploring the lab's divisions on LANL's late-1990s website. One called "nonproliferation and international security" caught her gaze.

"The fact that it coexisted at the lab with weapons design people," she begins, then pauses. "It's not like I went, 'Am I willing to work at a place that also does nuclear weapons?' I didn't really, because to me, the place was so big that I was like, 'Well, that's not my job, right?'"

It wasn't, and it hasn't been since. But the scientists working to modernize, update, and upgrade those nuclear weapons toil in buildings that—at least from that view on the ski hill—look awfully close to Light's office. While one group works on nonproliferation, the other keeps the weapons in existence.

Some feel those programs undermine nonproliferation work. Light, though, still sees her own contributions as helping keep the world safe. While space-based systems aren't currently tied into the Comprehensive Nuclear-Test-Ban Treaty Organization's monitoring networks, Light believes their data nudge things in the right direction: Having more squad cars makes catching speeders easier. And if the United States believes it can catch all (or almost all) speeders, it might be more likely to ratify the treaty, which right now it has only signed—a less-binding agreement. "If I could do any small thing to help the Comprehensive Nuclear-Test-Ban Treaty be ratified, I will," she says, specifying that this is her personal viewpoint and not the lab's.

The country currently can—and does—reserve the right to perform nuclear tests, if that's deemed scientifically and technologically necessary to make sure the weapons, you know, "always, never." In addition to wanting to reserve that right, politicians don't want to ratify until they know for sure they can tell if others are playing by the rules.

Light's work in detection could bolster support for the treaty by providing deeper insight into other countries' hypothetical nuclear detonations or lack thereof. She and collaborators are in the process of gaining even more insight, bringing together explosion data taken from the ground and from space. "Currently, each sensor does its thing, and experts from each phenomenology do their thing," she says.

In other words, determining whether a nuclear test (or explosion of war) has taken place isn't like in the movies, where someone essentially looks at a flashy red screen and identifies a nuclear explosion. Instead it's a tedious process in which the groups of scientists in charge of each type of sensor look at their data in isolation and then bring their conclusions together manually. While there is some automatic integration of data, detonation detection programs don't typically look at ground- and space-based information together. Seismic experts analyze data from sensors that pick up Earth's shaking. Acoustic experts dig into infrasound waves, noise too low for you to hear. Electromagnetic experts check out emissions of radio waves. Physicists study how radioactive atoms are distributed. Other scientists, like Light, help watch from space. Each group gives its individual analysis to others, who eventually rule on whether the physical signals represent a nuclear explosion or not.

In the view of many detonation-detection experts, including Light, that process is too human intensive, too slow. And that feeling is part of why, in 2021, Light and another LANL scientist, Josh Carmichael, joined forces. Carmichael has spent years working on software and mathematical analysis to detect explosions using sensors on the ground. Together, the pair got a grant from the lab's director to join their efforts, fusing ground- and space-based information into one software system that could automatically alert officials to potential nuclear explosions.

As above, so below. Or, as they call it, iNDD, which stands for Integrated Nuclear Detonation Detection.

Carmichael believes the same thing as Light about his work—that trying to make nuclear detonations more quickly and reliably detectable helps keep apocalypse at bay. "Everyone wants a world where they are certain or highly certain that they are safe and secure," he says later, "where we don't have to worry about annihilation."

That's why he and Light hope to show—over the three years their project will run—that mashing all that data together automatically allows scientists to pinpoint a boom more quickly and reliably detect events that take place more quietly. "I've been kind of bitching for a couple years that we don't already do this," Light says.

Instead of just bitching, she's moved on to making nuclear detection what it ought to be.

LANL IS WELL SUITED TO TAKE ON SPACE-BASED NUCLEAR DETONATION detection because, historically, it's overseen the development of the very exploding things it's keeping an eye out for. To put it in the weapons lab's own words, "It takes a weapons lab to find a weapons lab."

Light's research has roots back in the nuclear age's early days. In 1963 the United States signed the Limited Test-Ban Treaty, promising—along with the Soviet Union and the United Kingdom—not to set off nuclear weapons in space, under the sea, or in the atmosphere.

The United States had already been working on test detectors since 1959, like Vela from the Advanced Research Projects Agency (DARPA's predecessor). Los Alamos and Sandia had begun to work on the Vela satellite system later that same year.

Just a week after the treaty entered force, the United States launched the fruits of those labors. From then into the 1980s, various Velas kept eyes on Earth.

They were looking not at the kinds of things that show up in a picture of an explosion but at the energy that detonations release, most of which your eyes can't see. Just a couple milliseconds after an explosion, neutrons, gamma rays, and X-rays flood from ground zero and interact with the environment, producing more radiation, including regular light and electromagnetic pulses.

Some of the satellites' first big detections, though, weren't of Soviet malfeasance. Instead, the Velas found a mystery that originated much farther from the homeland. On July 2, 1967, a pair of satellites saw a burst of gamma rays that didn't match weapons' patterns: two peaks of energy rather than a flashy burst and then a fade. Scientists didn't know what they'd seen, except that it *wasn't* a nuclear weapon. Laser-focused as they were on said weapons, they put the data away. But more similar detections soon happened, bursting all across the sky—suggesting they were coming not from here but from *out there*, in space. Systems like Vela draw scientific discovery and national security close, and that makes sense: pure physicists and astronomers are often interested in the same data as weapons scientists because nuclear weapons are, in fact, just the physical power of the universe brought down to Earth.

Finally, in 1973, Los Alamos scientists published a paper announcing the discovery of what later came to be known as gamma ray bursts. Today data suggest they happen when a massive star goes supernova or when two neutron stars smash into each other. These are the universe's own bombs, way more effective than even the most modern nuclear weapon.

The Vela satellites provided other nonnuclear boons, revealing details of, for instance, how the solar wind works and how a sheet of plasma wraps around Earth. They also did their intended job, detecting forty-one nuclear tests before 1979, followed by the controversial "Vela Incident," which may or may not have been a secret test no one's ever taken credit for. Determining who detonated a weapon is a branch of nuclear work called *attribution*, which involves analysis of things like intelligence, forensics, and location information.

Los Alamos's ALEXIS and FORTE satellites continued in Vela's tradition, revealing a new lightning phenomenon that could help scientists predict when and where severe weather will occur. Light came aboard at LANL during the FORTE mission, which aimed to demonstrate detonation-detection technology.

The fruits of her current sensor project—now aboard the GPS constellation—haven't come easily. For one, the detectors have to be "rad-hard," or radiation-hardened, meaning resistant to errors caused by radiation in space. Rad-hard components always lag the state of the art in a kind of conservation of capability: if you make electronics better at resisting bursts, you take a trade-off in processing power. Since space computers are so good at functioning in a harsh environment, they're not as modern as what you might have on your desk.

"I'm always forced to use old systems," she says. "We're slowly coming into the twenty-first century."

PART OF THAT—HER OWN KIND OF MODERNIZATION EFFORT—INVOLVES her project with Carmichael, whose office door features stickers denoting underground nuclear tests the United States has performed: Disko Elm.

Bowie. Muleshoe. Amarillo. Whiteface. Distant Zenith. Back when they were performed, the Overton window—the range of policies deemed acceptable to the mainstream, which shifts over time—overlooked the nation's own explosions.

Every explosive event—a nuclear blast, a meteorite's breakup, an industrial accident—produces a plethora of signatures. Nothing is just a flash or a boom. It is a flash, and a boom, and a shock wave, and an electromagnetic outcry, and more. The idea that explosions sport multiple signatures came on strong when Carmichael saw firsthand that lightning didn't just illuminate the sky and make some noise: it also created intense radio waves.

One day, he was driving around with his stereo on as a New Mexico storm churned above him. And lo: whenever lightning flashed in the sky, static fuzzed out of his car's speakers—radio waves from the sky, caught on his car antenna.

If you combine the different signatures from a given explosion, they assemble themselves like puzzle pieces, together revealing a picture of the world as it actually is. Lightning's combined signature looks different from a meteorite's or a bomb's. But their individual components aren't always distinguishable when viewed in isolation.

As Carmichael contemplates this concept, in August 2021, he sits close to a canister of whey protein and a box of Mighty Leaf tea. His shelves are stocked with thrillers: *Fracture Mechanics*, *Geology of the USSR*, *Detonation*.

Looking at them, Carmichael talks about the *fusing* of data—bringing different types together. He then stops to seek a synonym. "'Fusion' has a different meaning in the nuclear world," he explains.

Carmichael seems to see everyday life in this fusible way too. He likes shooting rifles because each time you fire, sensory feedback of all sorts reveals the state of things. Metal on hands. Trigger resistance. Flash. Boom. Recoil. Reverb. Thwack.

He also likes to take his daughter for hikes every morning "so she can get multisignature information." Maybe it's the sight of deer legging along. The evergreen smell of a high-altitude morning. The feel of the strata in a mesa's rock.

One day when the pair ventured out, they ended up trapped between a mother bear, on one side of the trail, and her cub, on the other. The cub played cluelessly while its parent grew more and more agitated. Carmichael watched the bear pace back and forth, threatening him with her very existence. He was suitably deterred.

CARMICHAEL FIRST TESTED HIS OWN DATA-FUSION SYSTEM IN THE SUMmer of 2019, aiming to see whether it could determine whether a conventional explosion was in fact an explosion. He was at home when he saw news of a relevant tragic event: a vat of propane at the Philadelphia Energy Solutions oil refinery had caught fire outside the city. The blaze, visible from afar, soon burst outward in a huge dome of fire that saturated camera sensors. In videos of the explosion, nearly the whole screen goes white for a second, before the flash dies down and the roiling flames and smoke reveal themselves.

Seeing that boom, Carmichael fired up his laptop and began siphoning data from the likes of air and seismic sensors and social media. Soon enough, his software synthesized it all and told him what he already knew: there was a 99 percent chance that an explosion had occurred in Philadelphia.

It had worked.

"Because no one was hurt, I can say it was awesome," he says.

The Philly explosion was also a good test case for his new project with Light: a weather satellite picked the event up from space, giving them orbital information that they could feed into the model. Light and Carmichael can also practice their fusion on fires, volcanic explosions, and large meteors that explode in the atmosphere. From these natural events, the team can see if their algorithm can determine that, yes, there's been an explosion, but, no, it doesn't look like a nuclear weapon went off.

Or, maybe someday, it will look exactly like that.

# CHAPTER THREE

We must look reality in the eye and see the
world as it is, not as we wish it to be.

—2018 Nuclear Posture Review

The Integrated Nuclear Detonation Detection (iNDD) team's computational problem is complicated. And they are only able to dream that it might be solvable because of computing capabilities that, in a lot of ways, exist *because* of the nuclear program. To better understand its own bombs, the nuclear complex has long created computer simulations and models of those complex systems.

An even better digital understanding of nuclear weapons was required after 1992, when the United States did its last nuclear test. Four years later, the country signed the Comprehensive Nuclear-Test-Ban Treaty, promising—sort of—that tests wouldn't take place on its soil anymore. But tests had played an important role in weapons development, demonstrating that a device worked as planned—or failed to. The ability to simply flip a switch and get that answer was easier, in a lot of ways, than what scientists have to do now, which is to mimic weapons' behavior down to tiny detail, inside computers, and complete experiments with separated components—never the whole.

But as modernization, including pit production, goes forward, still-unanswered questions about weapons' behavior have come to the fore.

JILL HRUBY SITS CALMLY IN THE HOT SEAT AT HER CONFIRMATION HEARing, wearing a blazer and a tricolor necklace, a bottle of Dasani water glinting nearby. She'll soon speak about the country's nuclear future in a voice that typically drifts only a few notes below and above her baseline frequency. Testifying before a Senate committee, during a pandemic May day, she hopes to be named head of the National Nuclear Security Administration (NNSA), the organization of which she's been a part since 1983.

Senator Jim Inhofe kicks off Hruby's confirmation with a stark statement of American atomic standing. "We are more threatened today than we've ever been before," he proclaims. "We simply can't afford to fall further behind our adversaries." To avoid that fall, the country needs to bring its nuclear arsenal up to date and match or best the standard of other countries—at least, that's the implication he's setting Hruby up for.

NNSA is after all, and under Hruby's anticipated leadership, aiming to avoid a scenario in which its weapons aren't on a par with its competitors'. But the nuclear modernization program's rationale, Senator Angus King of Maine points out, remains opaque to the average citizen. "Many Americans don't understand why we're doing it," he says. "Those of us that have lived through prior decades have to realize the Cold War is in a thirty-year rearview mirror."

And yet nuclear weapons still reside, and almost certainly will continue to reside, at the center of military strategy. The bomb—the thousands of bombs—always pulse beneath the geopolitical surface, their threat to would-be aggressors keeping that aggression at bay, or so the thinking goes. "Our essential defense posture for the past seventy years has been deterrence of others' attacks on us. Is that not true?" King asks Hruby, beginning a long line of rhetorical questioning.

"That is true, Senator," she replies.

"And deterrence rests upon your adversary understanding that you have a capacity to make life very difficult—that is a euphemism—for them and the will to use it," he continues. "Is that not correct?"

"That is correct, Senator," says Hruby.

"And if your deterrent is out of date, in bad shape, not modernized, not able to be delivered, then that in fact makes the country less safe, does it not?"

He's speeding up his cadence, as he counts down to philosophical launch.

"That is correct, Senator," Hruby says, matching his velocity in magnitude and direction.

"And so the modernization that we're talking about is essential to maintaining the peace. Isn't that correct?"

"It is," Hruby affirms.

"I think this is important because people back home in Maine say, 'Why are we spending all this money on nuclear weapons?'" King says. "The reason is because we never want to have a nuclear war. The irony or the paradox of nuclear weapons is we build them so we'll never have to use them. And that strategy has in fact worked for over seventy years. Is that correct?"

"That's correct," says Hruby.

Together, these two political players are putting on a display of the deterrent ideal, performing for the others in the room, the official record, the assiduous watchers of C-SPAN, and curious leaders abroad the ritual of its affirmation. And key to the modernization Hruby will oversee is the production of new plutonium pits.

THOSE WHO WANT TO MAKE THE PITS SAY IT'S NECESSARY TO ENSURE that "always, never" holds. The pits in the arsenal are decades old and might either become dangerous or no longer pack the right punch. The country not only needs those plutonium pits, according to officials, but also needs to have the capability to make plutonium pits—which requires infrastructure and a specifically skilled workforce.

But whether pit production is a scientifically necessary endeavor, at least at this particular moment, remains unclear. NNSA maintains that new plutonium parts need to exist because aging makes them impure and possibly less reliable, and the current plan will replace all the pits in the entire arsenal before that becomes a problem.

The science of how plutonium ages isn't settled though. According to the secretive group JASON, which gives advice to the government, and a statement in the federal 2021 budget explanation, NNSA hasn't done enough research on how the plutonium will change as it gets older. (JASON, featuring a rotating set of prominent scientists, was created after the launch of Sputnik, the first satellite, to assist the government with defense issues; many of its reports are classified.)

NNSA's Michael Thompson, a principal assistant deputy administrator, doesn't think research is enough. "Experts disagree on where and when there will be any issues impacting performance with the pits currently in the stockpile—all agree they will not last indefinitely," he says, "and we cannot continue to try and study our way out of this concern." But whether or not they stand behind the need for pits, some critics doubt whether NNSA's production plan is safe, realistic, or necessary. On the first front, there's the environmental problems previous pit production caused. And Los Alamos's plutonium work will happen in an existing facility called PF-4—the site of numerous safety concerns in recent years. To the second point, the building where Savannah River will do its deeds was previously meant to alchemize weapons material into reactor fuel. But experts projected the project would run $13 billion over budget and thirty-two years behind schedule. Given the numbers, the project got canceled. NNSA is now repurposing its infrastructure into a pit plant.

On the question of whether new pits are necessary, a 2006 JASON report suggested the existing pits have lifetimes of at least one hundred years. Internal research from Lawrence Livermore National Laboratory suggests the cores have lifespans of at least 150 years. But a later report, which JASON released in 2019, told a different story: the advisory scientists suggested the Department of Energy (DOE) should restart pit production "as expeditiously as possible" to "mitigate against potential risks posed by [plutonium] aging." In the end, it seems, they agreed with Thompson.

Still, the plutonium pits' birthdays aren't the only concern driving production: most of the first pits won't actually go toward replacing those in stockpiled weapons; they'll get plopped into modernized ones, some with different requirements, instead. Given all that, some see pit manufacturing as enabling *better* bombs. Rather than dulling the arms race, they could be upping the ante.

With classified information involved, it's impossible for members of the American public—whose taxes fund pit production and who deal with the global consequences of the US military's actions (along, of course, with the rest of the world)—to evaluate the options and to parse expert evaluations. Things are moving forward anyway though. And the towns that host these twenty-first-century Rocky Flats will have to deal with what that brings—more money, more jobs, more risk, a firmer place in the nuclear weapons–oriented modern world. And the labs that, for decades, have worked somewhat abstractly on "always, never" "safe, secure, and reliable" will have to reckon with what it means to take a more active role in weapons development and production.

Many employees are philosophically onboard with that because of the belief that global stability rests on a blue-ribbon deterrent. But that isn't *just* a strategic philosophy. It is, like most things, also about money. "The Pentagon will sometimes come out and say, 'Our weapons are the best in the world. They can return anything. No country should even try and do anything crazy,'" says Geoff Wilson, director of the Center for Defense Information at the Project on Government Oversight. "And then they'll turn around and be like, 'All of our weapons are falling apart.'" We have to replace them with better ones ASAP, and that will require big budget lines, in other words.

In the course of Hruby's hearing, she points to the deterrent importance of producing those plutonium pits. "This is the biggest issue I think facing NNSA today," she tells the senators. Scientists know, she says, that the behavior of pits shifts over time. "And the planned pit production program allows us to get ahead of that and make sure we don't get to a point where we need to test the weapons in an underground—in a nuclear explosive—test to make sure they're reliable," she says.

Coming from Hruby's mouth, the statement sounds almost like a threat: unless pit production goes forward, testing might. During the past few years, officials have in fact discussed whether the United States should spool up the full-scale nuclear detonations—something it hasn't done in decades and currently has a voluntary moratorium on.

A moratorium the country could lift if it proclaims it must.

A LOT OF MONEY, HISTORICALLY, HAS GONE TO A LOT OF ATOMICALLY explosive experiments. After the initial 1945 Trinity test of the world's first atomic bomb, detonations continued. Many of these explosions happened at a desolate facility not far from Sin City, a place today called the Nevada National Security Site (NNSS) and formerly known as the Nevada Test Site.

It's about an hour's drive from downtown Las Vegas, along a long, straight road that also passes by Creech Air Force Base. The exit ramp curves into the test site's entrance but doesn't tell you the road is for authorized personnel only, as some military exits do. In fact, it doesn't say that this is the road into NNSS at all. It claims only that you're on the way to Mercury, the word printed on a green road sign like the name of any other town, a blue rectangle beneath the name warning that there are "no services."

But the issue isn't that no gas stations exist. If you pull off the highway, you will only be able to drive a couple hundred feet before signs scream at you that this is a restricted area and you had best get out of there. No gate blocks your arrival in Mercury, which is the name an Atomic Energy Commission representative initially provided to the US Postal Service when establishing the site's base camp. Today, the site is near the Desert National Wildlife Refuge. But one hundred million years ago, the animal species here were different, and so was the landscape. Liquid—an ocean—sloshed over the solid surface beneath.

Back then, it's almost certain no being on this planet had considered the concept of a "bomb." The planet itself, though, had its own power, just like the stars do: Submerged tectonic plates slammed together and sent mountains shooting upward. Volcanoes started to spout ash and steam. When their hot material cooled, it formed a kind of conglomerate-looking rock called *tuff*, the fusion of different ingredients. That rock was then itself lifted into different mountains, arched into basins. Later, ice ages broke up and eroded the land, creating the gravelly and sandy slopes of modern times—times when Joshua trees poke their speared heads and bodies out of the flat places and the bare mountains look both like they were pushed up and like they might fall back down.

Not long ago, when the landscape looked as it does now, which it only will for a blink of cosmic time, men came to this searing desert as part of a project called Nutmeg. Nutmeg's goal was to find a place in the continental United States where scientists could detonate the strange and powerful, semicontained stars they'd built on Earth. The land outside Vegas seemed suitable.

This part of Nevada could handle the explosions and requisite infrastructure year-round. The land, then part of the Las Vegas Bombing and Gunnery Range, was already owned by the government. The mountains that time and circumstances had thrust up created an amount of privacy, and it all provided some radiological protection for those who might find themselves nearby, whether in downwind towns or the big city of Las Vegas.

At the time of its founding in 1950, the test site covered 680 square miles. It's more than doubled in size since then and is now larger than the state of Rhode Island. Whereas the NNSS used to offer limited tours, its perimeter became an impenetrable border at the start of the pandemic and didn't reopen for years.

During the site's lockdown, the public could perhaps best experience it from the sparkling city of Las Vegas and its National Atomic Testing Museum. Sponsored by the likes of Lockheed Martin and Northrop Grumman, the museum tracks the history both of NNSS and of Sin City's relationship to it.

In the early days of the test site's existence, Las Vegas business owners worried, reasonably, that blowing up apocalypse machines just outside their gambling establishments might cut down on tourism. But the atomic explosions, visible from the city and brighter than its lights, in fact became their own attraction. They were, like Vegas itself is, a spectacle.

The town started to actively advertise itself as "Atomic City" and to print schedules for bomb detonations, detailing the chronology of the booms that happened about every three weeks. Bars served atomic cocktails, and mushroom clouds rose up on souvenirs. People hosted bomb parties—food, drinks, music, conversation, a weapon that vaporizes whatever's nearby. "You could tell friends and family that your party was a blast, or that it bombed," as one informational card in the museum put it. There were Miss Atomic Bomb beauty contests, for which women created nuclear weapon–themed costumes. One winner, Lee Merlin, is enshrined forever in the museum in the form of a statue. Her outfit was a cotton mushroom cloud sewn onto the front of her swimsuit, the stem starting at her groin and opening into the cap at her breasts.

Even the un-fun parts of the weapons had a touch of joke to them, in Vegas memory. One postcard shows a photographer capturing a be-stemmed blast in the distance, with the text "This Will Be the Best Picture I've Ev—"

But even back then, the bombs—or at least other people's bombs—weren't just a spectacle. A binder of photos in the museum shows before and after shots of manikins who had the misfortune of living their inanimate lives in a fake town constructed inside the Nevada Test Site. The Atomic Energy Commission bombed it to show what would happen if someone bombed your town and why you should practice "civil defense": the somewhat futile act of protecting yourself against an atomic blast. "They could have been you," reads an ad from the time.

BETWEEN 1946 AND 1958, NEARLY TWO HUNDRED WEAPONS TESTS TOOK place in the Pacific and in the desert flats of the Nevada Test Site. After that came a few-year hiatus, followed by a barrage of new detonations. Operation Sunbeam's final four explosions were the last to take place aboveground. The United States' final atmospheric boom reverberated in 1962 from a Davy Crockett, a weapon named appropriately for the king of the wild frontier. But by 1963, the Limited Test-Ban Treaty had pushed all detonations deep underground.

When officials stopped testing in the atmosphere, there was no visual component for visitors, who then had to spend more money at celebrity shows (Barbra Streisand, for instance, was a surprise hit the year atmospheric experiments stopped) instead of staring at the sky show. But although no longer visual, the booms continued. The museum displays the giant drills used to dig the tunnels in which underground tests took place—bombs exploding beneath the Earth's surface, making the world tremble. They look like three-mouthed monsters, multiple rows of teeth in each maw.

When those tests happen, trillions of atoms react, whipping up huge pressures and temperatures, vaporizing the weapon itself and the rock nearby. What's left is an empty space, a cavity full of superheated vapor. The remaining rock crushes together and fractures, sending a shock wave out like an earthquake does. It then decompresses, and the vapor transforms back into a liquid. That juice slurps to the bottom of the cavity and solidifies.

Overhead, weakened rock falls, collapsing between the explosion's depth and ground level. That leaves what's called a *subsidence crater,* a ring of sunken earth that looks more like it belongs on the moon, humans having this time changed the landscape and rendered their home someplace alien.

The tests and the previous aboveground ones—colloquially called *shots*—aimed to illuminate, in swirling and smoky color, aspects of the bombs' safety, their design, their effects on the environment and humans, the ways material might disperse in an accident, and the physical signs the United States could use to detect others' detonations—as people like Tess Light and Josh Carmichael are seeking to do. They were also, though, aiming to enable a better comprehension of nuclear physics itself. Sure, the weapons "worked." But how, exactly? Scientists didn't fully understand then. You may be surprised to learn they still don't completely get it today: much of NNSA's modern work continues to chase that elusive how, the inner workings of the fundamental materials and forces of the universe.

From the 1960s through the early 1990s, the United States continued its nascent quest for understanding, performing 760 subterranean tests. These occurred not just at the Nevada Test Site and nearby Nellis Air Force Range but also—for slightly different purposes—near regular towns like Grand Valley and Rifle, Colorado, and Carlsbad, New Mexico.

Part of a project called Plowshare, the shots in those spots attempted to repurpose nuclear weapons for peaceful uses, like extracting natural gas and generating electricity. The program's name descends from the biblical phrase "beating swords into plowshares," or turning military technologies to civilian benefit. "Nation shall not lift up sword against nation, neither shall they learn war any more," continues the passage from the book of Isaiah.

Such a future, though, did not come into being back then; nor has it emerged since.

By 1992, the year after the Cold War ended, the United States had overseen 1,054 nuclear tests. Those experiments ended then, and a few years later the country signed the Comprehensive Nuclear-Test-Ban Treaty, putting a (somewhat tenuous) halt to all tests regardless of where they took place relative to the ground.

THE COMPREHENSIVE NUCLEAR-TEST-BAN TREATY changed the nuclear complex drastically. For their entire history, weapons scientists had been able to observe an explosion, judge that it had happened as it was supposed to, and so infer that the bomb was behaving as *it* was supposed to. Should a test go awry, they would know they had a fix in front of them. "Nuclear testing was a wonderful tool," reads a 2014 article in the magazine *National Security Science*, published by Los Alamos. "It was also the world's biggest shortcut. It meant that we didn't have to understand all the details of a nuclear weapon and how it functioned."

They needed only to understand that it *did* function, then replicate the design and manufacturing, and—voilà—they'd have a bunch of working weapons. But eventually the scientists still needed to ensure weapons' good behavior and potency without tests. In the days of testing, weapons development and maintenance were like determining whether a cell phone works by simply turning it on. Once explosions were banned, scientists essentially had to understand whether every resistor, capacitor, and chip of the phone worked on its own, as well as how they worked together, and infer whether that meant the device would receive texts and run TikTok. For that, they had to understand precisely how and why these creations of theirs functioned. To that "always, never" end, the DOE established a program called *stockpile stewardship*.

The euphemism has a kind of peaceful denotation: we usually use the word *stewardship* when talking about things like environmental conservation or Christianity. In the context of nuclear weapons, though, it means inspecting and testing weapons parts, doing basic research on nuclear science, simulating nuclear detonations on computers, performing bomb maintenance, and modernizing both the weapons themselves and the infrastructure that supports them. It is about taking care of the weapons, as if they were members of a congregation, flock, or forest.

It sounds nice enough. But "stockpile stewardship" was not nearly as attractive an idea to the average weapons scientist as setting off a bomb was. As such, the labs in general looked askance at the test ban. "The opposition from the laboratories centered on a new argument to the effect that under a total test ban, the continued reliability of old stockpiled weapons (or of slightly modified versions of old designs) could no longer be guaranteed since it would not be possible to test these weapons to resolve any suspected problems that might have arisen," wrote anthropologist Hugh Gusterson in his book *Nuclear Rites*, an ethnography of scientists at Lawrence Livermore National Laboratory.

But the issue was perhaps as much about the humans involved as about the weapons' well-being: if they couldn't do tests, nuclear researchers would cease to be *scientists* in the same way. Sure, they could make digital models, comb through old data, and upgrade aging warheads. But "they would be unable to test any new theories and ideas. They would be experimentalists without experiments," wrote Gusterson.

The test ban thus also created a crisis of identity.

Today, however, many professionals who work at the tri-labs never had that earlier identity: they never took part in testing, the practice that used to be the centerpiece of the profession. They have only ever been stewards, looking after nuclear weapons as they edge into what the 2020 *Nuclear Matters Handbook* calls a third distinct era of nuclear times. The first era extended from the bombing on Hiroshima to the end of nuclear testing. From 1992 to 2018, the second era focused on sustaining the arsenal without detonations. But this new period involves an atomic revamp—and weapons production—without testing.

This is the modern modernization era, the era in which we find ourselves and that the scientists are adjusting to.

MOVING THROUGH THAT NEW ERA WILL COST HUNDREDS OF BILLIONS and perhaps multiple trillions of dollars. Ask inside the nuclear complex, and you'll generally hear that spending—approved and laid out as part of Congress's National Defense Authorization Act—is necessary. "If you look at some of the weapons that we've had out there, there's technologies that I think were created before I was born," says Rita Gonzalez, nuclear deterrence modernization director at Sandia. "And I'm not that young."

Electronic parts might need upgrades, like your TV would. Explosives don't have infinite lifetimes and can be swapped out for newer materials that are less likely to go off if, say, you accidentally drop them from a plane. It takes around ten years, according to NNSA's Thompson, to extend the life of a warhead through modernization.

The modernization program includes upgrades to the missile, plane, and submarine side of the nuclear complex too—although that is the purview of the Department of Defense and not the Department of Energy or the National Nuclear Security Administration.

The Project on Government Oversight's Wilson is skeptical of the need, versus the desire, for this level of modernization. "Because it's the United States' program, we only want the blue-ribbon version of these, right?" he says. "We only want the best of the best."

The idea is that if our arsenal ("the Deterrent," as defense types like to say) doesn't match or best the biggest, best capabilities out there, it won't actually deter attacks. And if the United States cannot deter, for its own benefit and that of its allies, even more countries may develop nuclear weapons to protect themselves. By building its stash back better, the country is actually keeping the world safe from acute conflict and freer from long-term proliferation.

But does the United States really need the best, or the most, or more to put the world in nuclear stasis? "If you believe in deterrence, you still have more than enough to get the job done," says Wilson.

While nuclear matters don't occupy the everyday minds of those who don't deal with deterrence every day, the people who *do* are the ones who run the world. Nuclear subtext underlies every tense international interaction.

But in such modern nuclear discussions, Wilson believes, the philosophical options have narrowed: you're either a hawk or a dove; no reasonable in-between exists. "Do you support the troops or don't you?" Wilson mimics. "Do you think the United States should defend itself or don't you? And if you're not for nuclear weapons, then you're for unilateral disarmament."

That limits political, academic, and public discussion of nuclear weapons policy, of programs like the controversial pit production plan, and of the wisdom and morality of a stockpile stewardship initiative dubbed *subcritical experiments*.

SUBCRITICAL DETONATIONS CROUCH JUST BELOW THE COMPREHENSIVE Nuclear-Test-Ban Treaty's threshold. They are scaled small enough that they do not fulfill the definition of "nuclear test." A full-scale weapon sets off a nuclear chain reaction, an unstoppable set of events that results in a mushroom cloud and supercritical fission and fusion. Nuclear explosive tests' subcousins do not reach such self-sustaining stature but can nevertheless hint at what is brewing inside the aging US arsenal, without plucking out a bomb and pressing "detonate." The goal is to see how different aspects of weapons behave—in other words, to see whether a phone is turning on and sending texts as planned, whether they're getting garbled or not going through, or whether the phone is frying its own circuit board.

During tests, if something about the weapon doesn't meet specifications, a "significant finding notification" goes out to designers, who then deem some notifications worthy of a "significant finding investigation." Between

1995 and 2005, designers opened 156 such investigations, of which 81 merited actions—like changing a material or component or the weapons' storage. Seven of the investigations were related to plutonium pits.

In the early subcritical testing years, the Department of Energy notified the public forty-eight hours ahead of every shot, then sent out a press release when the implosion was over. The alerts about such endeavors stopped in 2010, when the agency kept quiet ahead of a test named Bacchus, after the god of wine and general fun. Soon after, the press releases dried up too. In June 2011, NNSA began simply giving quarterly summaries of its setups.

And so it was that news of a 2017 test called Vega came not through any kind of "did it!" announcement but from the pages of an obscure newsletter called *Stockpile Stewardship Quarterly*.

The test happened at the NNSS, in a facility called the U1a Complex. This facility is a chimera: in the 1980s, engineers built a shaft for a 1990s nuclear test. In 2004, they shoveled out another shaft. Today's U1a Complex is the setup that spreads horizontally between the two. Its "tunnels and alcoves," according to NNSA, together cover 1.4 miles of subterranean space.

On the day of Vega, a weapons-grade, scale model of a plutonium pit sat more than nine hundred feet underground, surrounded by chemical explosives. The substance—a new and safer sort of combustible, called an *insensitive high explosive*—banged off, crushing the plutonium. As those radioactive atoms drew closer together, they started to split.

That sounds showy, but subcritical explosions actually happen in "containment vessels."

In 2019, the publication *Stewardship Science Today* replaced *Stockpile Stewardship Quarterly*. But the best place to find out about subcritical experiments is each fiscal year's Stockpile Stewardship and Management Plan, created for Congress.

According to those documents, the complex has kept busy since Vega, with subcritical experiment series like Sierra Nevada, Red Sage, and Nimble. Excalibur, meanwhile, is meant to qualify a new tool that lets the labs research plutonium aging, manufacturing innovations, and future life-extension plans.

And U1a is set to expand and update, again, with a program called Enhanced Capabilities for Subcritical Experiments, at a cost of more than $1 billion and a focus on plutonium pits, of whose behavior scientists will be able to make X-ray images.

Grab your popcorn: it's a cinematic future.


WHILE THESE SUBCRITICAL TESTS ADHERE TO THE LETTER OF THE Comprehensive Nuclear-Test-Ban Treaty, some say they may bump up against its spirit—especially since they happen in an underground facility that *could* someday support a big-time explosion. It might make other countries edgy that the United States could, in a matter of months, convert that facility back into a *real* test site.

As more than forty House representatives wrote in a letter of concern just before the very first subcritical test, Rebound, in 1997, "The U.S. is unwisely creating a testing norm under which other nations could justify conducting similar underground nuclear weapons experiments at their test sites."

Their worries were perhaps not unfounded. Causality is a hard thing to establish in both science and politics, but Russia and China have also done such experiments since Rebound. In April 2018, North Korea copped to them too.

Currently, the Nevada National Security Site has a readiness plan to become a full-on boom town again, able to host critical nuclear detonations. And it might someday: the country's true commitment to not testing has always been soft. Given the difficulty of maintaining and modernizing an aging arsenal, the 2020 *Nuclear Matters Handbook* notes that "uncertainties and challenges . . . may make it necessary in the future to resume some level of nuclear explosive testing to certify the aging nuclear stockpile." If the country resumed nuclear testing, it would set a dangerous international precedent, setting the stage for other atomic nations to do the same—and potentially setting off a chain reaction of similar tests. And detonating weapons, even in underground experiments, can lead to environmental contamination and health consequences for humans, as it did in the past.

"Certification" of the stockpile happens annually, when the directors of the three NNSA labs, the secretary of energy, the administrator of NNSA, and the head of US Strategic Command assess the safety, security, reliability, and effectiveness of the existing weapons—and determine whether it's possible to be *sure* about those qualities without testing.

Hruby, who was confirmed in July 2021 and thus is now a key part of stockpile certifications, held the line in her initial hearing. "My life's work has been in the missions of the Department of Energy and NNSA. Making sure we have a safe, secure, reliable stockpile is the foundation of NNSA, and I do not believe we need to test," she told the gathered, besuited politicians, "at this time." Her statement leaves room, as do official policies of the government for which she works, for future nuclear explosive tests and their fallout.

# CHAPTER FOUR

In real life mistakes are likely to be ir-
revocable. Computer simulation, however,
makes it economically practical to make
mistakes on purpose. If you are astute,
therefore, you can learn much more than
they cost. Furthermore, if you are at all
discreet, no one but you need ever know
you made a mistake.

—John Mcleod and John Osborn, *Natural Automata and Useful Simulations*

TODAY, BOTH SMALL-SCALE EXPERIMENTS AND COMPUTER SIMULATIONS underlie stockpile certification. To see if a weapon will work, scientists and engineers need to be able to use computers to predict and model the behavior of those devices down to tiny detail and make sure they line up with real-world results. Digital representations of the bombs—those of today and those of the future—are still beset by uncertainties, though, and full of surprises. And the surprises have increased as computers have gotten better and allowed scientists to see what they never actually understood in the first place.

It's perhaps no surprise, then, that a Los Alamos National Laboratory (LANL) employee wants to make nuclear simulations closer to nuclear reality. Outside a Theoretical Division building on Los Alamos's campus, Mark Paris sits at a picnic table holding a peach. He glances at two colleagues who're posted up across the courtyard. A tree swishes in the wind. Paris turns his gaze toward these things pointedly and then trains it on the conversation.

"We will not soon, it pains me to say, be free of the yoke of nuclear weapons," he says. He delivers this dramatic statement in a deadpan voice that trails off at the end of the sentence. It's the way of Paris, a goateed guy with an artist's spirit animating his physicist's veneer, who has a flair for the quietly, wryly flowery. But he works at Los Alamos anyway, helping the lab understand the fundamental physical behavior of the tiny particles within nuclear weapons.

Paris has long been fascinated by the idea that humans can use math and computers to replicate—and predict—reality. The obsession began in his preteen years, when he and his father were sitting in front of an Atari game console. Coding, naturally. The keypad was such that Paris would have to push hard to register his entries, only to insert five of the same character when he pressed the key with slightly too much strength. He also helped his father do theoretical calculations that showed how a graphite-fabric device would function, and they held up in the real world. "I was blown away that you could sit down with pen and paper and do this," he says. "That was it."

He was, clearly, already a theoretical physicist in the making, obsessed with the idea that equations—variables, constants, relationships—could explain how the world works, and how it would work if you pressed fast-forward.

Scientists in fields like Paris's, those who study the fundamentals of the universe—how everything came to exist, on the smallest and earliest levels—tend to have more philosophical relationships to the cosmos and the small stretch of it that includes Earth. Other sorts of physicists often get lost in the weeds—obsessed with the obscure details of this or that kind of star or deciphering the composition of this or that atmosphere for their own sakes rather than as an explicit part of the larger picture. Perhaps their field of study is what turned them pensive, or perhaps their pensiveness led them to this sort of research in the first place.

For Paris, at least, the latter path seems most likely. In college, he enrolled in some philosophy courses and was particularly taken by the book *Essays in Radical Empiricism*, a collection of writings from the philosopher William James. "He's basically interpreting the world in terms of the current state of the world," explains Paris, which is a better explanation of "radical empiricism" than you'll find online.

In a radical empiricist framework, the only worthwhile things on which to base your worldview are everyday experiences and the relationships between them. "To be radical," wrote James, "an empiricism must neither admit into its constructions any element that is not directly experienced, nor exclude from them any element that is directly experienced. For such a philosophy, the relations that connect experiences must themselves be experienced relations, and any kind of relation experienced must be accounted as 'real' as anything else in the system."

A young Paris was interested in gaining an empirical understanding of the world before he moved into a metaphysical one. But there was just one problem: "In empirical understanding, a lot of the things that were considered cutting-edge philosophy were BS," he explains later.

He was, instead, attracted to the different extreme: mathematical descriptions of the universe. "It doesn't matter what you say. It doesn't matter how you speculate. It doesn't matter what interpretational picture you have," he continues. "If the numbers aren't right, it's wrong. There's something you can sink your teeth into. There's a backbone."

"During this process, I became disillusioned with trying to reconcile these two things," Paris says.

Later, he would move on to contemplating the early universe in a somewhat philosophical, but also physical, way. "The ultimate question for me is origin story or creation," says Paris. If you wind back the universe's film and end up at the big moment of bang, the movie—the cosmos itself—just doesn't make sense. Where did the universe come from? "Causality breaks down," he says. "You have to say it was always here, which is annoying, or someone created it, but who?"

What he means, he clarifies later, is that before the big bang, in the universe's initial singularity, there is perhaps no causality. "If you don't have causality there, you don't have it anywhere. You can't have a description of the universe that's causal now and not causal then," he says. "And so my concern ultimately is that we don't have a causal universe now."

Put in layman's terms, he's worried the universe doesn't make fundamental sense in the way our brains make sense out of the universe.

Knowing this conundrum had no solutions he could access, Paris gave up. "There's no answers here," he recalls thinking. "I was on a quest, and I think I saw the quest has no terminus."

Attracted to the quantification of fundamental physics, though, he leaned hard into it. Initially, and for years. But lately, he's leaned harder into applying that physics knowledge to weapons-simulation software. Maybe he could make the details of those atomics and subatomics add up. Given that such weapons will likely continue existing, Paris thinks scientists should have a better grasp on the physics they can teach us. Maybe with that wicked knowledge, future humans can avoid the pitfalls that we of the modern moment and our predecessors have fallen into.

It is unclear, though, how, or if this is just the rationalization of a conflicted mind. Or one that just wants to keep digging into hard problems and needs some justification.

Nevertheless, Paris persists in this work. In his recent research, he's hit back on a hundred-year-old problem—one that scientists have been essentially glossing over since before the dawn of the atomic age. And one that even with today's supercomputers, they haven't really tried to solve. It goes back, says Paris, to the scientists who were just discovering what atoms actually are.

BY THE EARLY 1900S, PHYSICIST J. J. THOMSON HAD DISCOVERED THE electron and posited what came to be known as the "plum pudding" model of the atom: the electrons sit suspended in a positively charged sphere—like raisins in that beloved dessert. His contemporary Ernest Rutherford was also at work deciphering atomic structure. And he did not agree with Thomson's assessment.

Being an empiricist of a different sort, Rutherford devised the Gold Foil Experiment. In it, his team shot a beam of alpha particles—identical to helium nuclei—toward a piece of gold foil and watched what happened next. If atoms were pudding, with charge and mass spread evenly throughout, as Thomson suggested, the alpha particles should pass right on through, hardly deflected by the atoms.

But that's not what happened: Most of the particles passed directly through, shifting course only a small amount. But a small number of them shot off all over the place, some even hurtling back toward the experimenters. Something was deflecting them.

The implications might not seem immediately obvious. But the point is this: whatever the flying-off alpha particles hit must have only accounted for a tiny part of a given atom's volume, but that small volume—a sort of pit—must have housed a ton of its mass and electric charge.

Rutherford had, in this experiment, discovered nuclei—the ultradense centers of atoms—and the spaces between them.

Most of the world, it turns out, is the space between nuclei. We experience it as solid, but that experience belies its nothingness.

"You go like this," Paris says, hitting the picnic table, "and it's empty."

Even though the picnic table is in fact empty, his hand does not pass through the table. "It's bizarre to me," he says.

Paris has been chasing a strange thing not in Rutherford's atomic structure but in his math. When scientists calculate the distribution of scattered particles, they run into a problem: there's a singularity in the math, an infinity that doesn't make physical sense, when you look at particles scattered in the forward direction. Something is *wrong* in the math. It is not a true reflection of reality.

The scientists who work on nuclear codes, though, have been using the equation as is and simply discarding that not-true answer. This means that their code doesn't accurately reflect what's going on with the weapons. They have been ignoring a key strangeness in their model. In large part it's because everyone else does—because no one knows what's actually right and everyone assumes what they're doing is good enough. "This happens a lot: You read it in a book. Someone asks why you believe it. You say, 'Because I read it,'" says Paris. "Did you check it out?"

He pauses. "It doesn't work," he says, referring to the singularity lurking in the equation. "I checked it out."

Paris does a lot of checking out. He's also, for instance, been checking out the ways nuclear simulations deal with charged particles in general. For years, Paris says, the lab has focused on how neutrons—which have no charge—move around, carrying energy and momentum with them. But charged particles haven't received the same attention, even though they bump into things a lot more often than neutrons do, in cached weapons and exploding ones. "What they did for many years up to today is say, 'A charged particle is created and deposits its energy where it's created,'" says Paris. But the charged particles don't sit tight: they influence the other particles and atoms inside weapons, in ways scientists currently don't understand, because the simulations ignore that fact.

This too is an old problem: Manhattan Project physicist Hans Bethe knew about it back in 1935.

Using estimations like the ones in Rutherford's and Bethe's calculations, because the real deal is both difficult and unknown, means scientists' models don't mimic reality. The resulting errors and uncertainties in the output, from different approximations, might compound in ways that are hard to predict, making the whole less accurate than the sum of its parts. And that's potentially consequential when you're talking about nuclear weapons simulations, because it means the simulations won't precisely and accurately reflect reality—or, at least, that you can't know that they do or how they differ.

Trusting a simulation and approximations is different in nuclear weapons than it is when you're studying, say, a supernova. "Stakes matter," says Eric Winsberg, a philosopher at the University of South Florida and the author of *Science in the Age of Computer Simulation*. "If you're establishing the safety of a vaccine or that a batch of belt buckles is not defective, the quality of evidence that you need in one case might be higher than the other because your assessment of the seriousness of being wrong is different."

Nuclear weapons simulations are serious things, to be sure. And yet, even in such high-stakes situations, estimations or approximating techniques can become commonplace in situations where they shouldn't be. "People may come to accept a technique in a domain when being wrong is not that bad," says Winsberg. For weapons, that acceptance might occur because a technique is common in basic physics research, in which scientists are just trying to learn the fundamentals of the universe. And then they might go on to work on bombs, take an iffy technique with them, and use it to help understand the nuclear stockpile. "The standard for accepting [a technique] might be this high," Winsberg says, gesturing to indicate the height. "And then it becomes so commonplace that people apply it in a context where the standard ought to be *this* high." This high is higher.

In other words, norms can take over, even when the stakes are different.

Paris believes that's essentially what's happening when weapons physicists use approximations like the ones that he's recently been investigating. "Your code is not going to be quite right," says Paris. "How not right, I don't know. I think maybe 50 percent wrong, 100 percent wrong."

That amount of wrongness was perhaps OK for a time, when we didn't have to understand nuclear weapons as well, and also when they weren't quite so old. "It gets you a certain distance down the road," says Paris. But today's scientists at least ostensibly hope to push the simulations to the point where they reflect the world as it is, not a simplified version that is as convenient as they want it to be.

Paris isn't sure that will happen, though, within the current lab culture. It will only happen "when there's more people who see the trajectory between where we are and where we want to be," he says. "I would not say that is this institution."

Where he wants to see motion, he instead senses inertia. And that bothers him: in general, special drive and curiosity impel physicists forward—they're left sleepless because they are happily kept up by particle problems they can't solve. "Physicists are demented," Paris says. "Particularly theorists." And while he witnesses that kind of neuron-gripping creativity in individual scientists, he doesn't really see it holistically, as part of the Los Alamos culture. "I don't know if we have that kind of a zeal at the lab," he says.

He pauses and looks up at the tree above him, which stands solidly. He's about to continue talking about the lab's current culture but purses his mouth instead. "I see an Everettian split," he says.

Here's what that means: In certain physics frameworks, the universe is actually part of a multiverse, a whole set of universes. Whenever an event in one universe has multiple potential courses, that universe splits into two (or three or four). All of the options happen—just in different universes.

"In one universe, I say something sensible," he says. "In another, I'm getting fired."

Disappointingly, he chooses the former and obliquely quotes a former LANL leader, leaving his own meaning to inference. He tells me that director Peter Carruthers once said,

> Ever since I've been here, there's been an increasing trend, both externally and internally, towards the illusion that you can manage science, whereas all you can really do is to get good people who are interested in the subject you want to develop. This increasing accountability at all levels of the federal establishment exudes a cold air that drives out the kind of neurotic and creative people that you need to make a breakthrough. There has to be a feeling of freedom and reward. You can't get good science out of people who recognize that they are being managed.

Paris's reiteration of the quote suggests that he thinks the lab, today, is still trying to manage science and scientists.

THE KINDS OF CODES THAT PARIS IS CONCERNED ABOUT PLAY AN IMPORTant role in pit production. Los Alamos's Nathan DeBardeleben—whose LinkedIn description of his work used to say, simply, "Wouldn't you like to know?"—tries to bring that computational world to plutonium pits, among other objects. "I look at lots of crap we manufacture, look at the data that comes off of those things, and look for systematic defects," he says. "And we're pretty good at knowing 'this is good, this is bad,' but we're not very good, historically, at knowing why this is good or this is bad."

These days, it's important for him, and others, to figure out what's good and what's bad and why about the pits. The details of his work, though, remain as vague as his LinkedIn page. "A lot of stuff I feel like I'm not OK to talk about," he says. "I don't really understand entirely what I'm allowed to say."

It's a relatively common refrain among those with security clearances, a sort of corollary to the phrase "if you can't say anything nice, don't say anything at all": if you can't remember what's classified and what's not, keep your mouth shut. In the minds of the people with the secrets, after all, the knowledge is just knowledge, the secret and the public fusing into a whole that's not always psychologically separable.

But, anyway, DeBardeleben *can* say that computer engineers haven't looked at the data in this way before. That's actually a potential flaw of the lab, a defect in the secrecy setup that he himself is perpetuating with his air of mystery. "I was here almost twenty years, and I had meetings as recently as last week where I didn't understand how some of the things they were doing were done," he says. "That's vague, but it's classified stuff."

Those stovepipes and silos, as defense types like to call the compartmentalization of information, aren't the only frustration hardware and digital experts have in the nuclear complex. "It's very easy to come in here as a computing professional and say, 'Well, the right thing to do is this, we shouldn't be programming in Fortran, there's new programming languages, they're very modern, this is dumb' or 'You know, that's why it performs poorly,'" he says. "But you can't turn a battleship on a dime."

DeBardeleben compares working at LANL, a lab with the world's fastest supercomputers, to working with one hand tied behind his back. He doesn't mind the constraint.

A FEW STATES AWAY, AT LAWRENCE LIVERMORE NATIONAL LABORATORY, computational physicist Arjun Gambhir feels a combination of the issues DeBardeleben and Paris do: the simulations Gambhir works on are now so good that problems scientists didn't even know about are popping out. "We've progressed to the point where those aspects really start to matter," he says. In Gambhir's experience, having more computing power has actually created more questions than it's answered, and he's skeptical that the faster computers of the future alone will be able to answer all the new ones that will arise. "I don't think that's really a fault of the computers or of the people working on them," he adds. It's the fault of the questions themselves. To (hopefully) figure their answers out, researchers will have to build better models and algorithms and pair them with those ultrapowerful supercomputers.

Gambhir's California colleagues agree, to an extent. "As they do more, they find out they know less," says Terri Quinn, associate program director for Livermore's computing program. "They find out all the things they knew less about. And so then they do more."

Quinn, like everyone else, is unable to go into specifics about what "less" means. But imagine, she says, that a simulation spits out something that simply looks weird. "It's like, 'Oh, my gosh, I've never seen that before,'" she says.

In the outputs from previous software, the results weren't as sharp, and anomalies were eventually smoothed out. Today, with the finer detail, scientists can see what's wrong. It's kind of like having a blurry picture of a dog whose face and paws you can't quite make out. It definitely looks like a Labrador, though, so you think, "Cute!" And then someone hands you a high-definition shot of the same dog, and you see that it has a tumor on its nose and an extra dew claw, and it's bleeding. "They find out 'Oh, we've been assuming this is OK,'" Quinn continues. And it's not.

Whenever those problematic anomalies do crop up, scientists start a new project for each. "There is a list of things that they want to cross off that they could get away with before," says Quinn. ("Get away" is phrasing Paris would like, in that it implies a bit of complacency.)

But getting away with those things, whatever they are, becomes more dubious once you know about them and as the weapons get older. "The margin's getting narrower, and you've got to sharpen your pencil," says Quinn, "to make sure it can write that the weapons are safe and secure."

Otherwise, of course, perhaps they are *not* safe and secure—or perhaps you'll have to blow one up in Nevada just to be certain.

# CHAPTER FIVE

The computer was born to solve problems
that did not exist before.

—Bill Gates

Terri Quinn lords over Lawrence Livermore National Laboratory's computing center. Crossing the spare building one May day in 2022, she logs her credentials so she can enter a huge white room with white floors, topped with humming racks of processors. Steel plates beneath them shore up their weight and provide seismic stability. As Quinn walks the grounds, the computers' fans blow her hair around.

When labs like Livermore need a new or better supercomputer, they don't invent it themselves: they go to private companies and find ways to get what they need while also ensuring its long-term use in the industry. "They like working with us because we give them advance notice of what people will ask for in the future," Quinn says of the commercial computing partners' attitudes. And that's been the case since early in the nuclear age.

The fission-only bomb the Manhattan Project produced was simple, at least compared to its successor, the hydrogen fusion bomb. Understanding and predicting atoms' merging, in addition to their splitting—as for the bombs that followed and for modern weapons—required crazy math that they couldn't do by hand.

Luckily, one of the Manhattan Project members, polymath John von Neumann, knew of a computer called the ENIAC, built at the University of Pennsylvania. It may have run more calculations in ten years than humans had, total, in the species' existence. Von Neumann told the Los Alamos scientists about the computer and soon started figuring out how to make ENIAC calculate whether a thermonuclear bomb—also known lovingly as an H-bomb, or a hell bomb—was feasible. But using the ENIAC was a challenge: to run every new model, someone had to move its wires around, and the computer didn't belong to the atomic scientists, so they couldn't just stick wires wherever they wanted whenever they wanted.

Von Neumann thus did what a guy like von Neumann does: he simply invented a new computer, called the Mathematical Analyzer, Numerical Integrator, and Computer. MANIAC, naturally. Finished in 1952, it weighed a thousand pounds. According to a Los Alamos history, MANIAC's operators became so intimate with the machine that "they could debug code by ear, using a radio to listen to the interference patterns the computer generated," like a composer parsing the chords of a symphony.

MANIAC crunched the numbers for Project Mike, which led to the first hydrogen bomb. In a 1952 test called Ivy Mike, the device exploded in a roiling dark tower topped by a deadly dome. The explosion destroyed an island in the Pacific, literally wiping it off the map and leaving only a crater, the absence of a place that used to be.

The lab's computing needs remained unsatiated, though, and so a year later, Los Alamos built MANIAC II, which operated for two decades and leased a machine from IBM. By the mid-1950s, buying equipment from big companies had become a better bet for the Atomic Energy Commission than building it from scratch.

But it was the synergy—or capitalistic fusion—of the commercial computing world and the bomb makers that worked best: if the nuke people could encourage computer development that regular, commercial outsiders would also have use for, the computer companies had a long-term incentive to help their atomic friends out. For instance, IBM collaborated with nuclear types and the National Security Agency on a computer named Stretch. After a couple commercial iterations, the technology powering Stretch became part of normal commercial machines.

Some consider Stretch the Original Supercomputer, but the most famous of the early nuclear processors was 1976's Cray, which looked like a place to rest at a futuristic carnival. Technicolor panels fit together into a tall C-shape, and a mod loveseat covered the power supply.

Supercomputers were officially sexy.

But they did no good if they had no code to run, and so nuclear scientists developed software to elucidate what happened in the nuclear tests they were conducting and to predict what would happen in untested bombs. Results from tests then fed back into the codes, proving, disproving, and improving them, in a self-sustaining and critical process of revision. By the 1980s, the software could simulate in 3-D.

That didn't make the codes totally representative of reality, though: they were mere estimations, with flaws, uncertainties, unknowns, and inconsistencies. And so when the nuclear test ban came along, so did a problem: if the scientists could *only* simulate, and not test, they couldn't be as sure of their results. More importantly, they couldn't be sure of what they were unsure about.

There's a target for what scientists need out of simulations, says Alex Larzelere, former director the Office of Strategic Computing at the Department of Energy (DOE). "The center of the bull's-eye is 'It's OK, we're confident it's OK,'" he says. The area outside the target is "It's not OK, but we're confident it's not OK, so we can spend a billion dollars fixing the problem."

The middle rings, where you don't know if it's OK at all, are the places no one wants to be.

When nuclear scientists could perform explosive tests, they could try to home in on the problematic parts of the bull's-eye. Without testing, how would they know their weapons worked and were safe? And how would they build anything that didn't already exist? "It really was test-based the way they did the design," says Larzelere. "All of a sudden the laboratories couldn't do what they'd been doing for fifty years. And so it really became a huge problem."

The problem wasn't just scientific, though: it was also financial. "The business model has changed, you and your testing are no longer in charge," says Larzelere. "They were really not happy."

With the transition from testing and nuclear development to the more passive stockpile stewardship, then, the weapons scientists needed much better simulations than they had. In 1995, the DOE set up the Accelerated Strategic Computing Initiative (ASCI) to create and enable those better versions. As before, the program had to give the computer industry something that was also useful to it, not just to the labs. After all, as *How to Win Friends and Influence People* lays out, making someone think an idea is theirs to start with and beneficial to them is the best way to get what you want. And so the DOE focused on supporting commercial development of smaller computer systems that could be linked together, like Legos, into something larger. The pint-sized versions could go to all the consumers who weren't trying to mimic hell bombs; the keg-sized ones could go to the DOE as supercomputers. "All of a sudden you had much more powerful computers being produced out there because there was a demand for them," says Larzelere.

The program set lofty goals that Larzelere says people didn't think were possible and were perhaps a "political sales job." The project hit its metrics, though, whether they were political or not. In fiscal year 2000, DOE scientists simulated and visualized the explosion of a nuclear weapon primary, the initial fission component of the explosion. A year later came the secondary, which begins its job after the primary and involves fusion, and then a full system. In 2003, they modeled how a radioactive environment would affect the electrical response of a nuclear weapons system. And in 2004, ASCI spit out code good enough to help certify the nuclear stockpile as safe and reliable—a digital process that's been in place ever since.

The simulations helped scientists understand results from earlier nuclear tests that had previously left them scratching their heads. "More than once, after viewing a new ASCI simulation result, weapons scientists have remarked, 'Oh, so that is how it works,'" Larzelere wrote in *Delivering Insight*, a history of the program.

"The first step is always incredulous," says Larzelere. "'That's not right. That's a mistake.' Then you have to go down to the next level: 'That doesn't make sense. What's going on?' That's where experiment and theory come in."

Today, the ASCI program has morphed into the Advanced Simulation and Computing (ASC) Program. New supercomputers have come and gone, their names flashy and impenetrable: Roadrunner, Cielo, Trinity. A next-generation computer, called Crossroads, will soon enough run simulations with "full physics," according to Los Alamos. But the work isn't done: the Department of Energy has, for instance, embarked on the Exascale Computing Project, which aims to make computers five times faster than 2020's most powerful supercomputer. The first exascale machine came online in 2022 and was 2.5 times faster than the second-quickest machine on Earth.

While some worry that the United States will not just blow out these digital weapons but someday return to nuclear testing, Miles Pomper, senior fellow at the James Martin Center for Nonproliferation Studies, thinks there's little will for that within the nuclear establishment. "If nothing else, the national labs have sunk so much into the systems they have an investment in it now," he says. "And I think that's part of the point. It was sort of a buyoff."

It's worked, in that the simulations, at the NNSA labs, are helping with the nuclear modernization program, revealing how newly developed or developing weapons and subsystems will perform.

It's a bit like building bombs *in silico*. "We're not going to go back to testing, and we're making some pretty big excursions from the existing stockpile," says Rob Neely, Livermore's weapons simulation and computing program director and its ASC executive.

Twenty years ago, people would have said you couldn't design a new weapon—as the country will be doing with a bomb called the W93, which doesn't exist yet—and put it into your to-be-used pile without testing. They also would have perhaps said you couldn't alter, modify, or modernize existing weapons to as significant a degree as the United States is without those explosions. But those were the people who flew from Livermore to the Nevada Test Site as their "commute." "People would just get on the airplane in the morning, go down, and blow shit up in the desert," Neely says. Those were the digital doubters. "It wasn't until they shook the earth that they really felt confident," he says. "Now, we've got a new generation of designers coming up. The codes are their test site."

TODAY, AT LIVERMORE, A COMPUTER CALLED SIERRA RUNS CLASSIFIED weapons simulations. Sierra doesn't just use traditional central processing units, like your laptop probably does. Instead, it also uses graphics processing units (GPUs) optimized for things like movies and video games. It's a modern example of lab scientists watching where the computing industry is going and trying to flow in the same direction. "What we're trying to do is stay mainstream, as close to mainstream as you can," Quinn says of the way the lab thinks of computing developments, "but make targeted investments."

Scientists, nuclear and otherwise, have thus appropriated GPU technology to make it useful for data processing, rejiggering their code to work on this nontraditional type of system. (On the side panel of one Sierra rack, for the record, Rick Perry's signature swoops. When he was Donald Trump's secretary of energy and visited the facility, he wanted to sign it, and so he did.)

Sierra will eventually be superseded by Livermore's next iteration of supercomputer, named El Capitan, after the sheer cliff face located about three and a half hours from the lab. Although El Cap is not complete yet, three testbed systems—baby beta versions of the final deal—already rank among the top two hundred fastest supercomputers in the world.

The actual El Cap will be the first exascale computer belonging to the National Nuclear Security Administration (NNSA) (the first such machine in the United States, called Frontier, is at Oak Ridge National Laboratory in Tennessee, which is overseen by the DOE but not NNSA) and will cost the agency around $600 million. The broader Exascale Computing Project, a collaboration between the DOE, NNSA, and industry, has, unsurprisingly, been a big investment, approaching $2 billion. The financial and the research and development focus on computing power has serious motivation. "That's because of some really daunting problems," says Quinn. "We'll probably never see a push like this again."

Neely, though, thinks that kind of pause is the right move: it's time to slow down and focus on the codes themselves, not the processors. "Let's not throw yet another exotic computer at us for a while, so that we can think like mathematicians more than computer scientists," he says.

But Quinn is still interested in finding out what different sorts of processors might add to the equations, if they were also repurposed to run nuclear codes, as GPUs were. Livermore, in May 2022, was trying out chips, from the companies Cerebras and Samba Nova, optimized for artificial intelligence and machine learning, since that's where industry is going. A few months later, NNSA announced it would be evaluating such chips for its "post-exascale" future.

"How do we go to the next level?" Quinn asks. "Because we've pulled the GPU card. There's not a big card left."

"Quantum, maybe," she muses. "But that doesn't seem to be real soon. So there's that."


THE DOE IS NEVERTHELESS PURSUING QUANTUM COMPUTERS, COMPUTERS themselves made of the same tiny components their simulations are attempting to understand. "One of the things that we know that a quantum computer is good at is simulating other quantum systems," says Andrew Baczewski, who works on the technology for Sandia. Perhaps, put philosophically, since only a bat can understand what it's truly like to be a bat, quantum computers can help scientists figure out how the quantum world works. (On the other hand, humans aren't very good at understanding humans, and so the analogy sort of breaks down.)

Instead of using the traditional *bits* that regular computers use—which can be set to either zero or one—quantum computers use *qubits*, or quantum particles that can also be set to some proportion of zero and one at the same time, in a state called *superposition*. No explanation of how quantum computers work is satisfying, because quantum mechanics doesn't make much sense to our radical empiricist brains. As physicist Richard Feynman, who famously studied the subject, said, "I think I can safely say that nobody understands quantum mechanics." But suffice it to say that a quantum computer can do more atom-relevant calculations much faster than a traditional computer.

The Department of Energy is betting on quantum computers, to the tune of many millions of dollars. And maybe they could do for today's nuclear simulations what old-school supercomputers did for earlier weapons scientists.

One of Livermore's quantum computing projects is called the Quantum Scientific Computing Open User Testbed (QSCOUT), which became operational in early 2021. Its goal was to help the DOE evaluate quantum prospects in the near term, give people hardware they might not otherwise have access to, and help build a roadmap toward a larger machine. QSCOUT's initial form was a three-qubit system that let university and corporate researchers have a go at their projects for free to figure out what worked, what didn't, what was introducing errors, why, and how to make the field of quantum computing better. By 2023, they were up to five qubits.

When researchers rent time on quantum computers from private companies, by contrast, they often have to pay, and they don't get the level of access that QSCOUT gives them gratis. "We're sort of a white-box system that people can really learn from," says Susan Clark, head of the project.

QSCOUT lives in a lab at Sandia where wires traverse the room and a vacuum chamber houses, among other things, complex optical components. There are several powerful lasers. But the high tech, here, sits in stark relief to the low tech: to keep dust out, systems are covered with what looks like clear plastic shower curtains.

While QSCOUT is geared more toward fundamental research and understanding, one of Baczewski's projects is geared toward ascertaining whether quantum computers could have solid stockpile stewardship applications. "But in order to actually build a large-scale quantum computer that would be useful to solve some intractable problem, we need it to be really, really, really good," says Baczewski. "Really, really, really good" is likely a long way off.

The lab is reaching toward that goal anyway because, on a classical computer, the more atoms you're simulating, the more memory you typically need, in an exponential fashion. "Quantum computers offer the prospect of circumventing that," says Baczewski. Quantum computers only need to get more powerful linearly, because every qubit actually adds an exponential new amount of power. The difference between classical and quantum computers is like needing to buy a new fridge, and then a bigger new fridge, when you want to add a couple items to your nutritional stockpile—versus being able to stash those two new blocks of cheese in a drawer of your original fridge.

But the quantum future will be difficult to realize, in part because the DOE and NNSA already have such powerful classical computers: besting them is no elemental task. And scientists have built their current codes for decades, says Baczewski, to *approximate* atomic and subatomic calculations when they don't have the bandwidth for the full-scale deal. The resulting solutions are "actually really, really, really good," he says.

More to the point, the existing approximations and computing hardware do work at least pretty well, whereas quantum computers don't really work at all yet, on the large scale. "The projects that I lead are dedicated to understanding at what point will there be basically a kind of a breakeven between the capabilities of our biggest and best quantum computer and our biggest and best classical computer," says Baczewski.

DOE may have spurred on the supercomputing industry, but Baczewski estimates that the organization and the nuclear complex aren't leading quantum commercial developments. "I think that, in some sense, we will be the user base," he says. Nuclear weapons scientists and the DOE will harness invisible particles to understand invisible particles and will perhaps someday reconcile the real world with their math.

Maybe.

# CHAPTER SIX

Those who today claim the treaty will end
nuclear testing once and for all will be
greatly shocked.

—Jeane Kirkpatrick, in a Senate hearing
on the Comprehensive Nuclear-Test-
Ban Treaty, October 7, 1999

ON FEBRUARY 24, 2022, AFTER MONTHS OF TROOP BUILDUP AND UNCER-
tainty about what was to come, Russia invaded Ukraine. A missile flew
into an apartment building. Bridges blew. People started living in sub-
ways. Ukrainian ballet dancers and computer programmers put on fa-
tigues and took up arms, civilians in the fight for the sovereignty of
their nation. The Ukrainian president, Volodymyr Zelenskyy, filmed
selfie-style videos on the street, unshaven. "We are all here defending our
independence, our country," he said, straight into the camera. "And it will
stay that way. Glory to the men and women defending us. Glory to the
heroes."

Soon after the initial invasion, a forty-mile-long convoy of Russian soldiers stood outside the capital city of Kyiv, poised for intimidation if not
yet attack. Some of their equipment was broken down, some sabotaged.

But the facts remained: Russia's military was much bigger than Ukraine's, and although the country was putting up a solid fight, if this became a war of attrition, it seemed Russia would surely prevail.

Russia also seemed fairly likely to prevail if no one came to help, and Ukraine was left alone, like a rogue planet in blank interstellar space—with only foreign Stinger missiles, volunteer citizen-soldiers from abroad, and frozen Russian bank accounts to keep it company.

Indeed, other nations didn't come to Ukraine's direct military aid. Western nations kept their actions indirect, in part because Ukraine wasn't a member of the North Atlantic Treaty Organization, which guarantees protection to member nations, and in part because of the possibility of a nuclear World War III.

In the early days of the war with Ukraine, Russian president Vladimir Putin placed some of the country's nuclear weapons on a higher-level alert than previously. In a speech, he warned the rest of the world to leave him to conduct his invasion in peace. He urged anyone thinking about trying to stop him to reconsider. "Russia will respond immediately," Putin said, "and the consequences will be such as you have never seen in your entire history."

The nuclear threat was so naked that even a fact-checker wouldn't have called it "veiled."

Six days later, the events pressed on the minds of Los Alamos scientists, who were generally unsure whether or how the ongoing and escalating conflict would affect both the world and their work. It was clear, though, that classical ideas about deterrence did not always neatly or nicely apply to this situation. Instead of being deterred from potentially starting a nuclear, world-sized war, Putin had instead used his nuclear arsenal to threaten nuclear action against anyone who interfered. The nukes did not act as a carrot for peace: they became a stick to beat others into acquiescing to his actions. The existence of the world's nuclear arsenals may have allowed unprovoked violence to continue, rather than stopping it, because in addition to not wanting Putin to use his nukes, the world didn't want to begin a retaliatory exchange.

Nukes have in fact been deterring a world war, making atomic states reluctant to intervene against another atomic state. In this instance, however, the nukes are also allowing a dictator to bully and cow some of the most powerful nations on the planet.

The term *nuclear compellence* could also be applied to this philosophical use of the weapons. "Nuclear deterrence refers to the use of nuclear threats to discourage an adversary from carrying out an unfavorable action," the authors of a 2019 *Strategic Studies Quarterly* article explain. "Nuclear compellence is the use of nuclear threats to persuade an adversary to carry out a favorable action."

In this case, the favorable action is a nonaction: other countries staying out of Ukraine.

That inertia hasn't left people at Los Alamos National Laboratory (LANL) totally calm about the potential for the war to go nuclear. In March 2022, aware of those possibilities, Tess Light and Josh Carmichael wondered if their Integrated Nuclear Detonation Detection (iNDD) system could play a role in the conflict, should the unimaginable happen. "I think that if we get our asses in gear and things go very badly, we might have something to bring to the table," said Light, sitting in her office across the table from Carmichael. There was blue ink on her hand, the letters smudged and fading, a message in the process of being lost. By "very badly," Light meant, of course, that someone—she didn't say on which side—might set off a nuclear weapon. Not as a test.

Light and Carmichael were about to call a special meeting to see if they could put together a sort of beta version of iNDD. "What we're doing might be super relevant in the near future," Carmichael agreed. He then waxed a bit philosophical about whether humans tend to evolve toward a more peaceful state or that's just a story we tell ourselves to feel superior to the people of the past.

"WWIII isn't an impossibility," he concluded. "It could be more probable than improbable."

LIGHT AND CARMICHAEL'S ABILITY TO CREATE iNDD 1.0 IS CONTINgent on the hardest part of any nuclear project: humans.

After the lab approved their original proposal, the two iNDD founders brought on other scientists to help them make the James Bond–like detection system a reality. But the newcomers aren't yet seeing their vision. They're stuck in the stovepipes of their own specialties—the very stovepipes this program aimed to tear down in the first place, in an effort to merge existing systems for better results. Instead of solving that compartmentalization problem, their team has so far been replicating it. Part of the issue, Light and Carmichael both believe, is that they've been trying to improve nuclear detonation detection by videoconference.

Everyone is sick of saving the world via Webex.

Another part of the problem, though, lies in modern scientific culture and scientific personalities. "Scientists are difficult to work with," Light, a scientist herself, says. "I often prefer to work with engineers."

Engineers like to take a problem, come up with a practical solution, and implement it, she says. Scientists, meanwhile, are often driven by their own personal curiosity. Satisfaction comes from the intellectual chase—not necessarily from producing something useful. "If something doesn't serve the purpose of your creative outlet, you're done," Carmichael agrees. "Scientists are selfish."

"They need handholding," Light continues. "They get into their thing. When you say, 'I need something practical,' scientists shut down."

Still, they're trying to push their team toward the practical and make them think like the iNDD algorithm (someday) will: together, toward an end.

That's not what modern science—or at least modern science at LANL—necessarily cultivates or rewards, so researchers aren't conditioned to work together on problems with real-world applications. In Light and Carmichael's view, the lab can get stuck in an earlier era of science: the realm of the solo genius, publishing papers alone, hoping to revolutionize physics with his (it was usually his) mind. In today's world, though, the biggest challenges—climate change, nuclear threats, pandemics—are interdisciplinary and not even strictly scientific. Breakthroughs almost never come from individuals, and most truly important topics have direct applications on Earth, not just a place in an academic journal.

Los Alamos has perhaps failed to fully adjust to that reality. "We think if you haven't changed a textbook, you're B-string," says Light.

"Anyone who says they're changing physics is full of shit," she adds. "The physics revolution was a hundred years ago. We have such a cult of genius in this place. LANL is in a 1950s bubble where they don't know the lone wolf researcher isn't it."

Either way, they're trying to get their new researchers to see all of that and the iNDD vision, even if their epiphanies have to come via video-conferencing software. Carmichael has come up with a criminal analogy to explain the vision to the team. Imagine, he says, putting his hands up to paint an air picture, a courtroom. A lawyer is presenting pieces of evidence: a broken window, tire tracks, witness testimony. On its own, each piece of data doesn't mean much. Tire tracks exist all over the place. Broken windows have nearly infinite causes. Witnesses can say whatever they want. Then a window expert and a tire expert come to the stand, explaining what sort of tread rolling at what speed could make such indentations and what the shard pattern means about the object that broke the glass. Any of the experts taken separately doesn't give a clear picture of what happened. But take the three pieces of data together, along with context from glass and tire experts, and you may get a guilty verdict.

In the nuclear world, swap the tire tracks for seismic data, the window for an optical burst, and the witnesses for infrasound waves. And swap the experts for an algorithm that knows what "nuclear" looks like in all those domains.

Voilà: via some tricky math, you can get a reasonable probability of whether some suspicious event was an atomic detonation—hopefully a lot faster than a court case gets a verdict.

Light likes a different analogy, likening iNDD instead to blind men who are each given one piece of an elephant—telling a variation of the standard parable. The man who gets the trunk, for instance, thinks it's a very large snake. But if they just talked to each other before they started analyzing the animal kingdom, they'd likely be able to tell they held parts of a pachyderm. "What we're really bringing to it is saying, 'Duh,'" says Light. As in, why wasn't it this way from the start?

There are, of course, reasons, like the human tendencies toward inertia and egotism and attraction to money. People like their individual expertise to stay relevant—the renowned researchers who understand how shock waves travel spent a lot of time learning that information. Engineers want people to be proud of their creations. They're often invested in maintaining the status quo, which rewards their particular expertise and technology individually.

That siloed technology is the result of how the world progressed after the United States bombed Hiroshima and Nagasaki. In the post–World War II era, the arms race led to satellites that could watch for nuclear detonations. Then nuclear testing dove underground, avoiding this panopticon. So scientists developed sensors that lived on or in the ground. In that evolution, the funding lines for the eyes in the sky and the ears on the ground decoupled. Today, both groups—and the subgroups within them—have their own separate sources of money. The human and programmatic separations have stayed that way: Isaac Newton's first law of motion, after all, is the one about inertia.

But maintaining those many siloed sensing and analysis systems has a price that, in Light's view, could tip toward an unfavorable cost-benefit ratio. If you fused the data as iNDD hopes to, you wouldn't need to make detectors more expensive, exquisite, and sensitive. You could rely on "good enough" sensors that, when their pings are put together, give a yes or no better than the more spendy sensors that operate in relative isolation. She believes working on iNDD could later provide an off-ramp into a budget-friendly paradigm that does the job even better than the current one.

ONCE THE iNDD TEAM BETTER UNDERSTANDS WHAT THEY ARE SUPposed to be doing, they'll move on (ideally) to doing it. That process will start with gathering all the data available about a list of big, aboveground booms that aren't nuclear explosions: space rocks exploding in the atmosphere, industrial blowups, lightning. Software will ingest many different data types—seismic, infrasound, optical, radio wave. It will look at the different data in concert and compare that full set—what people like Light and Carmichael call a *signature*—to that of an explosion. If the combos match well, the system will spit out a high-probability chance the event was a detonation.

Those nonnuclear events will form the baseline rate of false positives—things that resemble nuclear explosions but aren't.

Later, they will also ask the software to calculate the chance that those explosions were nuclear. Problematically, though, there isn't that much data on what signatures atomic events leave. "Nukes aren't going off all the time," says Light.

So after testing the system on the nonnuclear events, the team plans to use simulated nuclear tests to assess its results and figure out how often it misses an explosion: the false-negative rate. They will use software to mimic nuclear detonations of different sizes and shapes in different conditions and feed the resulting data into the fusion algorithm.

In all conditions—small bombs, big bombs, bombs in the atmosphere, in a storm, and on a clear day—iNDD should be able to give a reliable yes, when it's ready to be used in the real world. That's important because neither identifying something as a nuclear detonation when it's not, nor determining something is *not* a nuclear detonation when it *is*, is ideal.

Hopefully, a yes will not be required during Russia's war with Ukraine—in part because the system isn't yet ready, and in part because no one wants a bomb to go off. Unfortunately iNDD cannot predict the future. "Everything we do is called post-boom," says Light. It's a joke-toned term, producing the emotional distancing effect that joke-tones do.

"That's the only thing about my job that bothers me," Light confesses: her tool—which she hopes never to use—is most useful in a worst-case scenario.

ABOVE LOS ALAMOS, IN THE WINTER SEASON, PAJARITO LOOKS MUCH starker than it does in August. A wildfire years ago swept across the mountain, wiping out many evergreen trees that would otherwise color the landscape. What's left in those areas, and what's regrown, are often shrubby plants, leafless aspens, and other deciduous types that drop their leaves come winter. Old snow slowly melts into mud. Still packed high in some places, that snow is nevertheless icy and particulate, like pixelated mashed potatoes.

It's not hard to imagine the bendable trees up here flexing in the shock wave that accompanies a detonation—as seen in videos of two shots at the Nevada Test Site, where workers planted neat rows of pines. The trees stood tall and still, till—BOOM—a blast washed across the footage. The shock bent the trees forward as if in assault, and they stayed bowed, stunned, before bouncing back up, waving in the fallout.

Anyway, it could look like that up here if you were so inclined to view it that way.

ONE HUNDRED MILES UP THE ROAD, AT SANDIA NATIONAL LABORATO-ries, Nathan Michael, senior manager in proliferation detection programs, helps create the sorts of explosions that some detonation detectors seek. He's trying to figure out how blasts reverberate through different types of rock—to understand, like the two Los Alamos scientists hope to, how to better identify others' nuclear detonations, in this case, those underground, which don't show up in space. A novice would say, says Michael, "'You're just blowing stuff up.' And there's some truth in that notion."

He and his team have worked for around a decade on a project called the Source Physics Experiments. Although this particular experiment began a few years earlier, interest in this sort of work, it's clear, spooled up after North Korea tested a nuclear weapon underground in 2017, ripping off an aftershock just eight and a half minutes later, when the cavity the explosion had created collapsed. A series of smaller shakings came later. The United States wanted to make sure it could detect small, deep explosions in rock that might keep its own counsel. "If you're doing an underground test, you want the least detectable place and medium, right?" he says. So Michael and his team have been trying to determine how to see those more secret experiments.

Swapping nuclear material out for regular chemical explosives, they have blown up batches of nitromethane ten times at the Nevada National Security Site, in a borehole below Yucca Flat. Forty-two sensors have monitored and recorded the waves that rocked the earth, to see how this unnatural shaking looked and how it differed from a natural earthquake.

For one, explosions seem to let loose fewer large aftershocks than actual earthquakes do. And that's the kind of data the scientists plan to use to create software models of the seismic effects of nuclear weapons. It's the kind of information that could also someday feed into underground detection algorithms.

But the information that comes out of Michael's experiments could also someday undermine detection and monitoring work. If you know which rocks best hide explosions and precisely *how* small a detonation needs to be to evade detection, you hypothetically know how to hide your own nuclear experiments, should you want to do them. Like almost everything in the nuclear world, this work is two-faced.

# Part Two

# DUALLY NOTED

SOME SCIENTISTS STUDY THE CORES OF STARS OR, SAY, THE ATOMS THAT coalesced near the universe's starting line. Others try to understand electromagnetic signals from lightning or how earthquakes play out in seismic detail.

Some, meanwhile, study the cores of weapons or the atoms that form at the start of a nuclear detonation. Others try to understand how electromagnetism and the Earth's shaking can reveal a nuclear explosion.

At the National Nuclear Security Administration labs, many scientists live on both sides of that line, doing both fundamental science and weapons-related research. And that makes sense: there's often not much of a crevasse separating the two. Many methods, computer codes, and fancy instruments are equally relevant to basic research and what people around the nuclear weapons complex call "national security science" or "mission work."

How those two faces interact and how dual-use scientists deal are part of the United States' continuing nuclear course and its curvature.

# CHAPTER SEVEN

Plutonium has a quite extraordinary relationship with people. They made it, and
it kills them.

—Ian Hacking

WITHIN EACH LAB, YOU CAN FIND PEOPLE WHO THINK A FEW DIFFERENT ways about deterrence. Some, though not many, scientists and engineers are fans of the bomb: hawks, of a sort. Their philosophy skews Strange-lovean, the adjective describing people who have learned to really love the bomb.

But most people at the National Nuclear Security Administration (NNSA) facilities are neither hawks nor total doves: they don't favor immediate and total disarmament but seem to occupy a middle ground. They're people like Mark Paris and Tess Light, who believe that the while bombs should be phased out, you can't just flip a switch and have that happen. It takes time, and it takes their work.

A Type II middle ground, though, perhaps places a little more faith in the value of nuclear deterrence. These nuclear thinkers don't want anyone to actually *use* the nukes, but they do believe the existence of the weapons dampens large-scale conflict.

Joseph Martz, at Los Alamos National Laboratory (LANL), is somewhere in the middle of the middle ground. "I wonder if the activists on the outside understand that there are those of us on the inside that share many of their goals," he muses. Like Light, Martz favors eventually eliminating nuclear weapons. "That's not an extreme position here at all," he says of the lab.

Martz grew up in Los Alamos, the son of a statistician. And he reacted to his upbringing in a Newtonian way: he wanted nothing to do with Site Y as an adult.

But he would not, it turns out, escape LANL's gravity. His orbital capture happened the year he graduated high school, which coincided with LANL's fortieth anniversary and the final reunion of the Manhattan Project scientists. Physicist Richard Feynman, ever demanding, said he would only return for the gathering if the lab made sure kids were in the audience for lectures. "I was one of them," says Martz.

From the front row, Martz listened as famed physicist Hans Bethe spoke about stars' fusion. "It went over my head," Martz confesses. But Bethe's final words were about the bomb itself. "He said, literally, 'My generation created this challenge. And it is up to you to find a solution,'" says Martz.

Bethe's finger shot right toward Martz's forehead. *You.*

Despite this pointed gesture, Martz didn't think about Bethe's words much at the time. But the demand lodged somewhere in his brain, dogging him throughout college. "I kept waking up in the middle of the night," says Martz, "and kept hearing him."

Perhaps it's not surprising, then, that the local boy started working summer jobs at the lab. By graduate school, he was devouring historical records from the facility he'd vowed never to return to. He was particularly fond of stories of former lab director Norris Bradbury. "What Norris used to say was, 'We don't build nuclear weapons to kill people. We build them to buy time for our political leaders to find a better way,'" says Martz. "If the products of our work are ever again used in anger, we will have failed in our mission.'"

This semiclassical version of deterrence still vibrates Martz's core, decades later. But he thinks he has a path *toward* that better way. "The reason nuclear deterrence was so compelling to people coming out of World War II was because all other forms of deterrence had failed," says Martz. The awful technologies of World War I—gas, machine guns— had killed millions. Their threat had not stopped anyone. Treaties, diplomacy, defensive installations—nothing had worked. Then came the bomb. "In almost desperation, people said, 'Maybe this is finally enough,'" says Martz.

And maybe it has been: deterrence advocates like to cite a plot of death statistics from major wars, wielding it like libertarians pull out pocket Constitutions. Before the end of WWII, wartime casualties, as a percentage of world population, were high. In WWI, around 1 percent of people died. WWII, meanwhile, resulted in the death of over 2 percent of the global population. But since the bomb dropped, there has been no major world war that killed such a large number of people: the graph shows a steep drop-off in casualties since the end of WWII. Some point to said chart and say, simply, "See? Nukes work." In their view, the threat of the bomb has led to the reduced casualties and the lack of large-scale world wars. Yet Martz knows that the correlation does not imply causation: the existence of the bomb doesn't necessarily account for the decrease in deaths. "The question becomes 'Have we been lucky? Or have we been smart?'" he says. "And frankly, it's probably a little bit of both."

"I want to make sure we don't need the luck," he adds.

He believes the path away from luck may have begun with the nuclear test ban. "Nuclear testing was a tremendous tool," Martz says. "It was also the world's biggest shortcut." Given that these words almost exactly echo the aforementioned *National Security Science* article, they seem to be a mantra of a sort, like so much else in the nuclear establishment (either that, or Martz ghostwrote the piece).

Without testing, scientists had to work from the bottom up to understand all parts of their weapons. To validate whether the computer models actually spit out results that hold up, scientists like Martz work on experiments that cuddle up close to the not-a-nuclear-test line. In LANL's Dual-Axis Radiographic Hydrodynamic Test Facility, for instance, researchers blow up models of weapons' cores using surrogate materials instead of plutonium, take high-frequency pictures, and compare the outcomes to their digital predictions.

The resulting understanding of materials means that researchers can predict how components will behave together before they've even made them. The lab can then build those custom, complicated parts quickly using 3-D printing or the more formal "additive manufacturing."

This research creates what Martz calls a "capability-based deterrent." "If you as a country have the capability to construct and deploy nuclear weapons, is the capability itself a deterrent?" he asks. Maybe if other countries know you can quickly build nuclear weapons, you don't need an actual arsenal. Perhaps that threat would suffice. It's an argument Martz has made since the aughts, and one he's gotten in trouble for talking about publicly in the past.

For this capability-based deterrent to work, the NNSA labs need to be agile—quick to whip the potential into the kinetic. In that way, modernizing both weapons and the processes to build them is part of the way forward. "Can we build a pit in one hundred hours?" Martz asks. If so, we need fewer in hand.

It's a complex, and somewhat contradictory, argument: revamping nuclear weapons presents a path toward disarmament.

At the same time, though, the logic provides justification for the same programs that some say could *lead* to an arms race. Either way, conveniently, it keeps the lab in business.

Pit production, in particular, promises to keep LANL in *good* business, with around twenty-five hundred new people coming aboard and billions coming into the budget.

Martz is an expert in the substance that makes up these pits—and a sort of scientific fan. "If you had to describe plutonium in a single word, it would be 'complexity,'" he says. "In one form, it can be as soft and ductile as gold. In another, it can be as brittle as iron."

You can hear, in his words, his delight at its exoticism.

In the past, Martz studied ways to separate plutonium from other materials, like blasting it with plasma so it transforms into a gas, which can then be pumped out of whatever substance it's sitting in, separating the deadly from the not.

That work ended up being applicable to Rocky Flats. A couple years after the site's forced closure, element 94 was bursting from the containers left behind after the raid. "This alarmed a lot of people for the obvious reason," he says. Also alarming: the same thing also started happening to plutonium at Los Alamos.

One day at the lab, a worker emerged with a "tremendous" amount of plutonium on their gloves and clothing. Martz mobilized and got to work figuring out how the plutonium had gotten loose.

He went straight to the source. Unboxing that batch of element 94, he found that it had turned into a dense, yellow material, swelling and breaking through the inner packaging. The plastic enclosing it looked like it had been burned by some mischievous teenager with a cigarette lighter. "It was alarming," says Martz.

Then he adds, "It was interesting."

It was a wicked problem, exactly the type Martz felt drawn to.

Soon he discovered that the plutonium had also gained grams of weight. And he also realized the reason for both problems: When weather blew through, pressure changes pulsed through the packages, and they breathed a bit. Small amounts of plutonium snuck out and sat on top of the plastic packaging. Those changes ultimately made the packaging become brittle, break, and turn burn-brown. The plutonium's reaction with the plastic also created hydrogen gas, which wafted back in to react with the plutonium on the inside, forming plutonium hydride.

"That's bad news," says Martz: when plutonium hydride meets the air, it sparks.

On top of all that, when the packaging broke, oxygen also crept in and reacted with the hydrogen, which swelled the container.

None of this was good. "It happened to dozens of packages, not just at Los Alamos but around the country," he says. Humans had been, once again, ignorant of their own creations, unable to predict their futures.

Martz ended up writing the first formal standard for how to safely store plutonium, and he helped colleagues who were modifying plutonium parts for some of the last nuclear tests. "I didn't know at the time just what a valuable experience that would be," he says.

It was also, he admits, kind of fun. Others would never get the wicked opportunity.


WHEN THE LANL SCIENTISTS TALK ABOUT THEIR CAREERS, THEY often speak about such lucky breaks, cool coincidences, and fortuitous trajectories. It's almost as if an invisible force—what some might call fate—shaped the universe in ways that purposefully shaped them. Something inside of them feels, even if they don't *think* so, that the cosmos evolved in their favor, its variables interacting in beneficial ways.

That's how David Clark, director of LANL's National Security Education Center, talks about his own work cleaning up Rocky Flats, which he says is probably the thing he's most proud of.

At the time, he wasn't really ready to take that work on—he was, in academia's eyes, just a kid. But he was a kid who had written a definitive review article about how radioactive elements behave in the environment. And he'd done so at a fortuitous time: the Cold War had ended; underground nuclear testing had ceased; Rocky Flats had closed. People were talking about who'd make plutonium pits in the future—maybe, they thought, LANL. "Los Alamos had hired just about anybody that was any good from Rocky," says Clark.

But back at Rocky Flats, where they'd come from, there was cleanup to do—and scared residents and angry activists to contend with. "Plutonium is this mysterious, enigmatic thing that everybody's afraid of," says Clark.

To understand what exactly to fear and how much, the powers that be needed a specialized scientist. Those powers read Clark's paper, and then they gave him a call. "I'm thinking, 'I don't know that much. I'm a young guy,'" he recalls. But he said yes anyway.

Working as a team at Rocky Flats, he and a group of researchers immediately turned up a contradiction. Colorado's spring downpours often sent what Clark calls "a cappuccino" flooding across the site. During intense rainstorms, monitoring equipment registered plutonium at various locations across the site, including the Rocky Flats fence line. It was also in many of the groundwater monitoring wells the site had drilled. "And yet the people at the site said, 'Well, plutonium doesn't move,'" says Clark. It isn't soluble in water and so "should" have stayed put, not been carried along with the cappuccino.

That clear contradiction—it shouldn't be there, but it was—bred public distrust of the authorities.

Soon Clark and his team understood what was happening: The rain didn't dissolve the plutonium. It simply carried it, as particulates, like water carrying a rubber duck. Contamination in the well water, meanwhile, didn't mean the plutonium really ran that deep: site workers had just dragged it down from the surface when they'd made the well.

At the same time as his work at Rocky Flats, Clark began studying the aging plutonium in the nuclear stockpile. He was involved in NNSA's original study on how long pits would last, which JASON—that secretive group of scientists who advise the government—had reviewed. But he wasn't a fan of the group's conclusion: that pits would reliably last for a hundred years. "There were many, many issues associated with that," Clark says, "that probably aren't worth going into. Because it's just, you know, political."

But he does get into it a little bit.

"In science, you always have provisos and caveats," he says. "I was always taught as a scientist, never say never, never say always." (Except, some might say, when you're saying "always, never.")

"The JASONs filtered out all the caveats and said, 'Oh, pits will last a hundred years,'" he continues. "We never said that."


WHILE SCIENTISTS LIKE CLARK AND MARTZ KNOW A LOT MORE ABOUT plutonium than humans did when they first synthesized it, the element remains more mysterious than most. To preserve hard-won information about plutonium, Clark thus edited a seven-volume series called the *Plutonium Handbook*. "Hardly anyone's gonna read my stuff, probably," he says. "But I think it's passing on human knowledge."

Ignoring the unknowns—the things not yet in the handbook—can be dangerous, says Clark. Imagine, he continues, that an engineer has designed a bridge that won't collapse under commuters. "Then I come back and say, 'Well, guess what, the element that you built the bridge out of is radioactive,'" he says. This ingredient is constantly decaying into a different element. That new element is uranium, and its atoms, once created, zoom out with lots of energy and knock the atoms next to them out of their spots, creating defects, voids, and potential structural issues for the bridge.

In ten years, says Clark, every atom in the bridge "has found a new home."

*Would you trust that bridge enough to drive over it?* is the clear question.

*Do you trust the pits?* is the clear implication.

Clark says the aging of plutonium is "nonlinear," meaning that it doesn't degrade at the same steady rate, and Clark wouldn't be surprised if pits show future problems researchers didn't expect, much like the cigarette-lightered plutonium packages. "Which is fascinating and great if you're a plutonium scientist," he says. "But it does color the kind of advice that we should be making to our government." He and others at Los Alamos contend that the best hedge against that uncertainty—including the unknown unknowns, the scary part of the aforementioned bull's-eye—is to simply make new pits.

He doesn't, though, think that Los Alamos should be doing so itself. "We're a research facility," Clark says. "We're not a manufacturing facility." They would be better off developing the new processes for making pits and teaching them to others—somewhere else.

"But that's not my paygrade," he adds.

# CHAPTER EIGHT

Man can still shape his destiny in the
nuclear age—and learn to live as brothers.
Toward that goal—the day when the world
moves out of the night of war into the
light of sanity and security—I solemnly
pledge the resources, the resolve, and the
unrelenting efforts of the people of the
United States and their Government.

—Lyndon Johnson, on signing the
Nuclear Nonproliferation Treaty

SANDIA NATIONAL LABORATORIES SEEMS TO BE FULL OF TYPE II MIDDLE-
grounders: peaceful people who nevertheless hold nuclear deterrence in
their hearts. Sandia has a more engineering-centric culture than Los Alamos, and, maybe as a result, scientists tend to talk in more black-and-
white terms, less philosophically. Sometimes, in fact, those working on
nuclear programs don't think about the bigger picture much at all. For a
while, Tina Hernandez was like that.

In Spanish, *sandia* means *watermelon*, Hernandez informs me. And
so Sandia National Laboratories sounds like a place to study fruit. She

gets a kick out of this—got a kick out of telling people when she was first hired. As she laughs, her earrings, shaped like peacocks, dangle in different directions.

When Hernandez, whose water bottle has a Barbell Babes sticker and a miniature New Mexico license plate reading "ALLBAD," first came to Sandia, she worked directly on the weapons, developing neutron generators that spool up weapons' nuclear reactions. Her work, and that of her colleagues, felt very focused. Everyone was heads-down on their small piece of the puzzle. Deliverables loomed; stakeholders chattered. The fact that the puzzle was a big picture didn't really occur to her, day to day.

Of course, she did get indoctrinated into deterrent ideals the same way most people do: with that plot of wartime casualties, plopped into a manager's PowerPoint. "I was like, 'Wow, OK, I never thought about it from that perspective,'" she says. "Because I think when you tell people, 'I work on weapons,' they think you're just this warlord."

She does, she admits jokingly, give off a warlord vibe. "But then looking at it from that perspective of the deterrence and how we have avoided so many conflicts because nuclear weapons exist," she says, "I was like, 'Wow, this is actually a great thing.'"

If Hernandez couldn't live in a world with zero nuclear weapons, and if maybe that wouldn't be such a great thing anyway, she felt comfortable working to ensure the existing weapons were safe, secure, reliable—and updated. "Do I feel comfortable driving a 1950s vehicle that probably doesn't have seatbelts?" she asks.

The question is rhetorical.

That moral dilemma satisfied, Hernandez continued to focus on her neutron-generating puzzle piece.

After a while, though, she started looking for a change. Parsing the internal listings, she found the Global Security Directorate. It sounded like she might get to travel abroad, which she loved.

Only when Hernandez, who's now a senior manager in that department, made the switch did she see the whole puzzle. Her new job involved working on international safeguards, the programs that aim to make sure plutonium and certain isotopes of uranium exist in the amounts they should and are not diverted for nonpeaceful purposes, and on arms control, programs to make sure that countries only have the number of deployed strategic nuclear weapons they report.

Hernandez's work involves a lot of talking to other nations, some of which have the bomb and some of which do not. Her group oversees the development of technologies that underpin the "verify" part of "trust but verify": they create the tools that can tell whether countries are diverting special nuclear material or deploying more strategic weapons than they're legally allowed to. "At the end of the day, if there's a negotiation tomorrow, and something is put on the table that the United States doesn't think that they can verify—because we've never thought about it before, we've never looked into it—they're not going to be able to accept it," says Hernandez's colleague Jay Brotz, who in 2021 was manager of nuclear verification. "So our job is really to help the United States be in a ready position for when there are negotiations."

Officials need to verify things like whether a nuclear explosion has taken place—the kind of thing Tess Light, Josh Carmichael, and Nathan Michael work on—whether radioactive materials have been tampered with, lost, or stolen, and how many warheads are deployed.

The Sandia scientists like to show off innovations that help with this task in what's called the Technology Training and Demonstration Area. It's essentially a museum exhibit hall, unclassified, where visitors from across the world can learn about the sorts of nuclear-verifying technology that's out there. One prominent display shows fake containers of nuclear material boasting the sort of tamper-indicating break-me-and-I-stain-you seals you might find on a gas pump or expensive clothing and glittery dust suspended in a clear epoxy that you can smear on a missile, because the smear's specific glint pattern can't be duplicated easily.

Arms control verification always involves being able to tell a real warhead from a convincing mockup and showing that you're not playing a shell game—moving bombs from place to place as they're being inspected, so the total count is incorrect. That means confirming the individuality of each weapon and its status as a real radioactive exploder.

Problematically, no country wants to give away precisely how its bombs are built, so inspectors can't just go pick them apart. Instead, Hernandez and Brotz's team has helped develop verification schemes where the inspecting side gets little information beyond "yep, that's the real deal."

One such method involves placing a special detector between a known weapon and a suspected weapon. The detector takes their radiation signatures and compares them. "If those objects are the same, like they have the same radiation pattern, the signal will always cancel," says Brotz. "And so the detector will only see random noise." No human person even needs to see the original signal at all. "Now, if the host puts two different objects there, then it will actually collect sensitive information," he adds. "So it incentivizes the host to actually do what they say that they're doing."

Brotz is proud of this work and sees both it and nuclear weapons as keeping the world safe and stable in different ways. But even people within Sandia don't always understand their division: they think people like Brotz and Hernandez are antinuclear, as if their work to stop the spread of nuclear weapons cuts against Sandia's work to help build them. "Like it's something that is the opposite to and working against the mission of most of our laboratory," says Brotz. "And I think that's not what we do."

But Hernandez's global-security initiatives have made her rethink some of her weapons work and consider how her previous role had helped perpetuate a world of haves and have-nots. Countries that already have nuclear weapons get to keep them; those without nuclear weapons aren't allowed to make them. The former always have more power. Inequity was baked in. She also saw how the nuclear modernization program might look provocative to everyone: to the haves, who'd promised to decrease their arsenals, and to the have-nots, who'd promised not to build any up.

Sure, the United States could say its updates and upgrades simply made the weapons safer, but did it look that way to the outside world? Did US modernization hurt Hernandez's attempts to convince nonnuclear states to stay that way? "There's so many different things that I didn't question when I was in nuclear weapons," she says, "or didn't even think about until I came to nonproliferation and saw the other side of the coin."

She still feels that her side of the coin is clean, if complicated. To prove it—to herself as well as to the international community—she cites the official six-stage weapons development process. Work in the "1" process involves designing a brand-new system; work in the 6 process simply maintains existing systems. "We always stay in the 6 process," Hernandez says. "We don't go back to 1. And so it's easy to tell non–nuclear weapon states, 'Well, we're not making new weapons, we're always in this maintenance phase.'"

But the W93, a new nuclear weapon, *is* in the early phases of its development. And some see modifications and alterations of existing weapons as problematic too. In 2021, for example, the *Washington Post* revealed that new sensors aboard the modified W88 warhead, completed that July, allow it to detonate at more precise times over difficult, hardened targets like bunkers. Such sensors are destined for both sub- and land-based missiles. It's a capability improvement.

The W88 isn't a new weapon: Los Alamos scientists originally designed it in the 1970s. But the new components—wires, sensors, and batteries— are part of its modernization. Ask those at the National Nuclear Security Administration what such modernization is for, and they'll repeat the "safe, secure, reliable" line almost every time. But ask critics outside about upgrades like this one, and they may say what Hans Kristensen, of the Federation of American Scientists, told the *Washington Post*: that the modifications get uncomfortably close to new designs.

# CHAPTER NINE

The aims of pure basic science, unlike those
of applied science, are neither fast-flowing
nor pragmatic. The quick harvest of applied science is the useable process, the
medicine, the machine. The shy fruit of
pure science is understanding.

—Lincoln Barnett, "The Meaning
of Einstein's New Theory"

ADVANCES IN WEAPONRY DON'T JUST COME FROM DIRECTLY SEEKING OUT
advances in weaponry: they also come from research in basic science,
seeking truth—or at least an approximation of it—about how the universe functions. That idea has lodged deep in the brain of computational
astrophysicist Christopher Fryer, who is a firm advocate of working on
open, regular science and quieter national security science.

Fryer grew up in Ridgecrest, California. The Naval Air Weapons Station China Lake, which did work for the Manhattan Project, is located
there. In his childhood home, Fryer promised himself three things: he'd
never again live in a small town that was in bed with a single industry,

he would not use computers for work, and he would not travel as much as his father.

That didn't quite work out. Today, Fryer lives in Los Alamos, a town almost totally dedicated to the lab, creates detailed computer simulations that pull multiple branches of physics together, and runs around to scientific meetings. "Every choice I made, it seemed like it was innocuous enough," he says, "but it ended up going to breaking my three promises."

That's why he tells his own kids to always remember this: "Your initial plans don't matter as much as you think."

If Fryer had a time-travel machine, he could have skipped back and bestowed that wisdom on early nuclear scientists, who sometimes thought their choices might influence the future differently than it did. They believed that they might retain more control over their creation and that they could see the future as it would be—not just as they wished it to be.

But these early nuclear scientists, like those of today, couldn't always simulate what was to come accurately and sometimes failed to predict the knock-on effects of their research. And there are pretty much always knock-on effects. Physicists' work on straight physics is rarely just work on straight physics. It often has applications—both planned, like the bomb itself, and unknown at the time of the experiment, like LASIK surgery, which spun out of the technology used by space vehicles and satellites to dock with each other. Questions, then, arise about how much responsibility these scientists bear for how their discoveries might be used in the future.

That's a long-standing debate. One perspective, taken by early hydrogen-bomb proponents Edward Teller and Lewis Strauss, held that "loyal scientists had a duty both to investigate the secrets of nature and to leave the decisions regarding the consequences of those findings to political leaders," writes historian Audra Wolfe in *Competing with the Soviets*. On the other side of the debate, some scientists maintain they must imagine how their work could be used offensively, even if the research and knowledge seem inoffensive at the time of their conception.

Atomic scientists often consider these philosophical questions more than the average researcher, even today, in large part because their field's early work produced an apocalyptic weapon. "My impression is the nuclear scientists have been really thoughtful about what the implications of their work are," says Wolfe—socially, politically, and economically. "What they have been less good at, in my opinion, is a realistic assessment of their own political power and their own position in the political system."

The Department of Energy has always been interested in those sorts of experiments and instruments that have both explicit national security utility and fundamental knowledge potential: this is an example of "dual-use" research, simultaneously helpful to both the people and the military, to encyclopedias and to weapons. That's a relatively simple realm for the National Nuclear Security Administration (NNSA) to occupy, because the same rules that govern the high-pressure, high-temperature states of nuclear weapons also apply across the universe—in supernovae and the centers of planets. A machine that reveals how stars' innards churn bears a striking resemblance to one that illuminates the workings of a bomb's core. Software that shows how the early cosmos created atoms also illuminates how subatomic particles whirl within a weapon. Research on actinide elements reveals information about how Earth formed, but also how to pinpoint the origin of stolen radioactive material.

The former examples are cases of basic, or pure, science, driven by curiosity and what a specific field would like to add to humanity's base of knowledge. "Applied science," meanwhile, says Wolfe, "is science that's driven by an attempt to solve a specific real-world problem." But the line between basic and applied research is often blurry, the distinction one more of name than kind.

The motivations behind so-called pure research are also often far from it: they're influenced by topical trends, scientists' competition for funding, their desire for professional fame, their commitment to a long-standing hypothesis, and even politics.

And basic science—especially during the Cold War—was relevant not just to fundamental truths about the universe but also to national security. Science, officials believed back then and still believe now, is a tool of soft power: its output can, the thinking goes, win hearts and minds to the American way (or at least complicity with American dominance) by demonstrating how innovative and advanced US society is. As Wolfe notes in *Competing with the Soviets*, "Science could be a carrot as well as a stick."

On the stick side, though, even basic science could continue to win the battle of the laboratories, its discoveries fueling military applications beyond just the bomb. And that has been at least somewhat true. Research into electromagnetism, for example, has assembled itself into directed-energy weapons that can blind or scramble other countries' satellites. The directed-energy research—applied science, at this point—can then continue on its own sticky trajectory.

But the arrow also points in the other direction. Military-focused science, which usually begins as applied work, leads to civilian applications. Nuclear power, after all, controversially emerged from atomic weapons. The Defense Department runs the GPS satellite constellation that fuels Google Maps and Tinder. Brain implants designed to quell soldiers' post-traumatic stress disorder also alleviate nonvets' drug-resistant depression.

No matter what the intention or moral cleanliness, then, some words from author Aldous Huxley remain true: "Pure science does not remain pure indefinitely," he said. "Sooner or later it is apt to turn into applied science and finally into technology. Theory modulates into industrial practice, knowledge becomes power, formulas and laboratory experiments undergo a metamorphosis, and emerge as the H-bomb."


TODAY, FRYER WORKS ON BOTH *MISSION SCIENCE*, THE TERM FOR NNSA's weapons management and national security endeavors, and science-science, studying similar physical processes that play out in the cosmos. He believes the best scientists at the lab work on both sides of that line, like he does, driven by detailed defense concerns and yet connected to advances in the outside world.

For Fryer, the two pursuits often collide, particularly when the obnoxious level of detail required for defense-related pursuits proves useful to scientists who previously made progress with more rudimentary calculations. In his experimental work at the lab, Fryer is currently studying how radiation flows and how shock waves result from its fast movement. His experiments start with clumps of material. He hits them with a front of radiation (not unlike a weather front) to see how they behave in the presence of that energy. Los Alamos National Laboratory (LANL) already has software codes that simulate such physics, and Fryer's experiments aim to see if that software accurately predicts—and so its creators have understood, on some level—what will transpire in the lab.

The things that happen in the early stages of supernova explosions are, it turns out, relevant to problems in stockpile stewardship. When massive stars reach the ends of their lives, they explode, going out with a bang that produces so much energy that these dying suns briefly outshine all the other stars in their galaxy—combined. Near the start of such an event, a shock wave travels through the star and then breaks out, sending more shocks along with it, and blasting out X-ray emissions—just as happens during a nuclear detonation.

"A guy comes to me and says, 'We're thinking about doing a mission to observe this supernova outburst,'" Fryer recalls.

Not just any guy, of course, but an astronomer who proposes space-based observatories to the National Aeronautics and Space Administration (NASA).

"Wait a minute," Fryer told the colleague.

The man had to wait thirty minutes, but in that sitcom-length period, Fryer took the model he'd been building for his weapons experiments and edited it to simulate a giant star at the end of its life—upping the scale from one centimeter to ten quintillion centimeters and looking at elements that are actually in stars, not at vanadium, as Fryer had been doing in the lab.

"It's the same physics," Fryer says.

Today, he and the colleague have been working together to design a NASA mission called the Shock Interaction Breakout Explorer, much of it based on LANL codes and experiments. "NASA just doesn't have the money to do that theory," Fryer says. "They're focused on building those instruments, but to really get all the information out of the observations, they need a theoretical understanding. And that's what the national labs do really well, is this kind of applied understanding of things."

Though NASA didn't fund it in the most recent round of selections, the group was asked to apply again—a fairly usual trajectory, even for missions that eventually do go to space. "The science case was almost entirely built here," Fryer says.

FRYER HAS HEARD THE NEGATIVE VIEW OF WORKING AT A LAB WHOSE research eventually emerged as the H-bomb. "Look at this evil thing this lab is doing because they're designing these weapons of mass destruction," he mimics. That sentiment is part of why he tells LANL job applicants that if they're going to come work for the national labs, they should have an answer to the question of why their work is important and not evil. The answer, though, isn't just for the aggressive others: it's also for the future employees' own peace of mind.

One answer may be strictly practical. Today, defense-related research funding far outpaces that from basic science agencies. In fiscal year 2022, the Department of Defense accounted for 41 percent of the government's research and development money. The Department of Health and Human Services got 26 percent. NASA got 8 percent. The National Science Foundation accounted for just 4 percent. That means topics with national security applications will almost inevitably get more funding than those that don't and so may progress faster. Military interests can thus skew the path of science—for better, for worse, for neutral—by pouring more cash into areas of defensive and offensive benefit. Just look at how investments, flowing into military problems, led to modern computing. "Moore's law doesn't exist in a vacuum, particularly the time part of Moore's law, right?" says Wolfe. "That that has to do with how much money you throw at a problem."

University scientists, competing for resources, thus sometimes take military money for their academic pursuits, not always thinking through why a defense or intelligence organization would want to study, say, plasma dynamics. One answer, of course, is that unknown future applications could flow from relatively fundamental research, just as nuclear weapons eventually emerged from Ernest Rutherford's Gold Foil Experiment.

Beyond that, though, the military is often interested in the opposite face of the dual-use coin, a second application that has little to do with the questions that compel the scientist. Take a scientist who studies how microgravity affects germs' virulence, hoping to keep future astronauts healthy. Along comes a defense research program interested in the same question: the study could allow the military to pinpoint the biochemistry of virulence, which would likely apply outside space, which would logically show them how to *create* a dangerous bug.

Everyone gets cake and eats it too.

Such defense-funded science tends to attract researchers who are in love with extremely hard problems. "Some people, particularly scientists who like this idea of pure science, are attracted by the intellectual challenge of some of this work," Wolfe says. "Sometimes the military is interested in problems that nobody else has the resources or the wherewithal to investigate."

Scientists of all stripes get caught up in curiosity about their own wicked problems and sometimes care more about their own pursuit of knowledge than they do about its consequences. That attitude goes back to the Manhattan Project. "I did not want the hydrogen bomb because it would kill more people," Edward Teller, who worked on the first H-bomb, famously said. "I wanted the hydrogen bomb because it was *new*. Because it was something we did not know, and could know. I am afraid of ignorance."

Fryer likes hard problems and is obsessed with pulling physics together so that computer simulations match laboratory experiments that eventually match the real world. But in that pursuit, Fryer has lately been bumping up against the same problem as Mark Paris: estimation, simplification, approximation. The science being not quite right. Fryer's particular concern is about equilibrium—or, rather, the lack of it.

It comes up in experiments that use a device called a *hohlraum*, essentially a gold-plated cylinder. Inside the hohlraum, physicists place a small amount of nuclear fuel. If you heat the hohlraum, radiation spreads through it evenly, heating the fuel sample. Sometimes its resulting blowup is a proxy—an analogy, essentially—for a small-scale nuclear explosion, and sometimes scientists use it to attempt fusion, the process of combining atoms and releasing energy in the process; sometimes both things are happening at once. But there's a problem: In hohlraum simulations, electrons sling around in a sort of soup, the negative particles colliding with each other. Digitally, when they hit, the fast ones get slower, and slow ones steal the fast ones' energy, so their speeds end up distributed evenly.

It's much simpler this way.

"And alas, that's not true," says Fryer. The timescales on which those reactions happen in real life—nanoseconds—are actually too fast for the electrons to sort themselves so neatly. "The methods that we typically use are no longer valid," Fryer says matter-of-factly. "And so we're going to have to change how we do our methods."

He doesn't just mean figuring out how to make these particular models work better: he's talking about a more revolutionary shift in mind-set—of the same sort Paris is shooting for. "I can't really prove this comment, that I say we have to rethink about how we do things," says Fryer. They are going to have to start thinking harder, he believes, and developing more precise ways of understanding, and replicating, the physics.

He's currently working with people who do care about the details of this problem—plasma physicists, who model every electron individually—to see if he can get their work to line up with that of the researchers who simulate things on a larger, holistic scale. But right now, the physicists can't simulate the larger-scale whole and its individual details, because that makes the codes run too slow. Fryer and his colleagues have to try to get as much detail as possible, while balancing the current realities of both the computers and the codes. What level of fidelity is good enough in a simulation?

The problem with those questions is that people's answers differ.

# CHAPTER TEN

"Bringing Star Power to Earth," read a
giant banner that workers in 2009 unfurled
on the newly inaugurated National Ignition
Facility. Over budget and behind schedule,
the construction had taken a decade.

—William Broad, "So Far Unfruitful,
Fusion Project Faces a Frugal Congress"

CHRISTOPHER FRYER'S HOHLRAUM CONCERNS RELATE IN PART TO AN
instrument over at Lawrence Livermore National Laboratory and experiments that take place there. The National Ignition Facility (NIF) is a
flagship National Nuclear Security Administration (NNSA) instrument,
even though it was completed years behind schedule and billions of dollars over budget. And although it took nearly fourteen years to live up to
the middle word in its name: *ignition*.

In this context, *ignition* refers to the giant machine's ability to release
more energy from fusion, the combining of atoms, than scientists pump
in, a feat the lab announced it had achieved in December 2022. The lab
likes to make a big deal about what this ignition means for fusion-based
nuclear power—the holy grail of the green-energy sector. But NIF isn't

a practical kind of power plant, economically or space-wise, unless your region likes to spend billions and set aside football fields' worth of land for your electricity-generating reactor. And the "ignition" of that December isn't quite as *eureka!* as it sounds, because it doesn't account for *all* the power scientists pumped in.

That's all kind of irrelevant, though, because creating clean energy isn't actually NIF's prime directive. NNSA didn't spend so much money building the facility so that people could have cheap electricity and humanity could combat climate change. That might be a nice side benefit someday and a nice PR play in the meantime. But in reality, NIF primarily exists to help with stockpile stewardship, providing a better understanding of the roiling innards of nuclear bombs. Conveniently, the device didn't need to reach ignition to be useful for that weapons work.

Here's how it works: At NIF, 192 of the world's most powerful laser beams focus inside the top and bottom of a hohlraum, which contains a tiny pellet filled with fusion fuel. The huge laser energy strikes the inside of the hohlraum. The powerful burst typically happens around four hundred times a year, although that number decreased during the pandemic, and each burst set is called a *shot*. For fusion, the resulting X-ray pulse causes the fuel capsule to implode, squishing its constituent atoms, ideally close and hard and hot enough to fuse together—and keep fusing. Like a power plant, like a star, like a bomb.

On a warm May 2022 day, Laura Berzak Hopkins walks down the hallway that overlooks NIF's bay of "beamlines," where the pulses from 192 laser beams travel. At the end of the line, many meters away, are manikins wearing hard hats and vests, to show anyone looking through this window just how big the laser bays are. On the wall opposite the window, slightly pixelated pictures of NIF's laser beams loom like alien goat pupils in that they are rectangular and a menacing red. In the laser bays below, leaden blue curtains cover the beamlines, as if for modesty. To keep radiation from the experiments from reaching the outside—the offices where people work—shield doors two to three feet thick isolate the space, like those of a serious bomb shelter. In the control room, on the other side of those doors, each desk has a big red button for emergency shutdown. A warning box can light up too, with levels indicating "Danger," "Low O2 leave area," "High deuterium-tritium leave area," "Caution," and "Safe."

Clearly, there's some hazard here. So before each shot, lights flash and sirens wail, warning of what's to come. But when the countdown clock hits zero, the only noise that enters the control room is a click, the power and sound being contained in the other room. Hopkins, who works at the facility, has advocated that they make some kind of artificial noise to commemorate the occasion: a doorbell, firecrackers, champagne corks popping—some outside evidence that an energetic event just took place. But for now, it's just *click*.

After the shot goes off, counts from detectors measuring the number of neutrons produced—a quantification of the fusion reactions that occurred—pour in. Workers use this number as a measure of both safety and success: the more fusion reactions, the bigger the number. They call the time during which these counts are high the *stay-out time*—as in, "stay out of the experimental facility because the environment is dangerous."

When Hopkins first heard about NIF at a conference, she was in thrall. "I remember thinking, 'Wow, this is a huge facility with incredible capabilities,'" she says. But a curious question came to her. "I remember asking the speaker, 'You know, I'm just a little confused. How does this not violate the Comprehensive Nuclear-Test-Ban Treaty?'"

NIF's work can, depending on whom you ask, tiptoe awfully close to a weapon detonation. Some organizations, like the Western States Legal Foundation, a nuclear-abolitionist group, agree. The foundation believes that stockpile stewardship as a whole entrenches US reliance on nuclear weapons and invites proliferation. They see this risk in particular in instruments like NIF, which pass nuclear knowledge to more of the world, allowing scientists to learn things helpful to designing new weapons, and which are also just a bit below testing of nuclear weapons.

Instead of dismissing Hopkins's question, officials sent her "a lot of paperwork." The paperwork pointed out that NIF does not create a *nuclear explosive yield*, the term for energy that bursts forth from booming a bomb, and so doesn't violate the treaty. The answer and its details satisfied

Hopkins. "This is a question I asked early in my graduate school studies because I wasn't aware of what NIF was, as opposed to a question I asked as an experienced scientist," she says, noting that she doesn't have doubts about the country's adherence to the treaty.

Today, obviously, Hopkins works on NIF and has crossed over from the energy domain, where she started, to weapons science research.

Predictably, Hopkins has heard all the rebuffs: she's gone to the dark side. "You just want to blow things up," she mimics.

"To be fair, I do like blowing things up," she clarifies.

HOPKINS'S FIRST WEAPONS WORK WAS IN SURVIVABILITY, INVESTIGATING how bombs and bomb parts hold up against neutrons, X-rays, and electromagnetic pulses, which would bombard them if they themselves were subjected to the output from a nuclear blast. Officials are always concerned with whether their nuclear weapons could still work after experiencing that kind of energy. For deterrence to work as intended, the country needs to be able to strike back after a nuclear attack.

While she'll say a little bit about what those experiments look like—shooting laser energy at a hohlraum filled with gas, so it emits X-rays that illuminate nearby weapons materials—she won't get too deep.

"We go into classified territory quickly," she says, evading details. But while there's secrecy here, the lab would actually like to speak loudly, clearly, and widely about the forest, if not the trees. "It's part of our national deterrence strategy to indicate to our real or perceived adversaries that our weapons are hardened against any type of nuclear encounter," she says, so that the indication reaches enemies who might read this book.

Deterrence, she says, is also part of why—beyond clean energy—NIF's fusion-energy research is important. Scientists can publish that unclassified energy work and present it at conferences, even in foreign countries. "It's the tip of the iceberg," she says. (Others might call it the tip of the spear.) Progress toward ignition, and now ignition itself, reveals the tiny tip, which then hints to those in the know at what NIF is capable of and made for.

Nuclear weapons scientists in China and Russia can look at the chip of ice and understand what's beneath, weapons-wise. "The bulk of the mass is under the surface," she says.

NIF ISN'T THE ONLY INSTRUMENT THAT DOES LARGE-SCALE, DUAL-USE research. Sandia has its own, named the Z Machine, or just Z for short. Its lead scientist, Dan Sinars, has long felt tethered to hardcore hardware like Z, rather than just the thoughts, equations, and codes many physicists favor. It's been true ever since he did a research internship as an undergraduate. "By the end of the summer, I had written a couple of Word documents describing some things," he recalls. Decades later, he still sounds disappointed, because another intern had spent those months constructing a scientific instrument—something you could touch and manipulate that would send physical evidence of its existence through nerves and eyes.

It didn't seem that important to Sinars right away. In fact, he thought the other intern was inefficient. Every day, the guy would realize he needed, say, a capacitor, hike across campus to get one, come back, sit down for five minutes, realize he needed a resistor, and go for another walk.

"I'm thinking to myself, 'Oh, you need to spend a few minutes planning,'" Sinars says. "But I was *only* doing planning, and I didn't actually *do* anything. And so by the end of the summer, he had built a working instrument, and I had written a couple of papers."

Sinars knew, then, that he didn't want to spend his life writing papers. He wanted to be more like the guy hunting for capacitors. And so when he graduated, he moved to Albuquerque to work at Sandia, a place where he could do bigger work on bigger machines than he could at a university—machines like Z, which had come online not long before.

Z takes up a mansion-sized space in the flatlands around Albuquerque. There, it uses powerful electrical currents to whip up proportionally powerful magnetic fields. The resulting electromagnetic forces then create high temperatures, high pressures, and a whole lot of X-rays.

Those powerful, radiative highs represent conditions that don't exist naturally on this planet. They're so extreme that they exist only in labs like this one or coiled inside nuclear weapons. Beyond Earth, though, these environmental conditions are plentiful: they're the extremes of stars, cosmic explosions, and the inner cores of worlds bigger than this one.

Z essentially makes far-off worlds accessible and real, whether those worlds are just Jupiter or a postboom Earth. The scientific knowledge to be gained from tinkering with those conditions excited the young Sinars. It was different from the science at his university—bigger, more consequential—and happened in a place with landscapes and mind-sets different from those back east. He could have a whole new life by coming here, where the sky seemed unimaginably large, its horizons too far away to really register. He would look at ominous clouds on their edges and think, "There's a thunderstorm. We'd better get inside." The storm wouldn't arrive for an hour. Where he came from, by the time you could see the weather, it was already on top of you.

Coming from academia, Sinars didn't understand what a machine like Z could do for nuclear weapons—science was, and is, number one for him. But understanding how it ties in with nuclear weapons increases his sense of purpose.

On an October 2021 day, Sinars walks into Z's "high bay," a room that he says smells like science—hydraulic oil and hot electronics. Sinars sniffs, looking down into a big tank. "If you drop anything in the oil section, let me know, because we will have to get it out," he says. "And please don't drop yourself."

Sandia's got a cautious, parental safety culture, more akin to a factory floor than a university lab, with signs regularly warning against slips, trips, and falls.

The showpiece of the room is a big cylinder, ten feet wide and twenty feet high, out of which you can pump all the air to make a vacuum. During an experiment, a target hangs out at the cylinder's center. These targets vary in their nature, depending on the experiment. Some are thread-spool-sized sets of tungsten wires, each five to ten times thinner than a human hair. "Other times they are fingernail-sized, flat panels of a material we are going to shock or compress," says Sinars. "And other times they are solid metal cylinders the diameter of a pencil containing fusion fuel." Z does a few types of experiments: some, for example, to understand the behavior of materials under extreme temperature, pressure, and radiation; some to understand fusion.

Transmission lines as wide as a car, insulated with water or the afore-mentioned oil, connect capacitors to the chamber.

When Z spools up, the huge capacitors charge for about 90 seconds. At the end of a countdown, they discharge all that electricity toward the center of the cylinder, in about 1.5 microseconds. For a brief period, Z pumps out sixteen times more power than the world's electricity plants put together. That power hits the targets, and the resulting forces cause them either to implode, in the case of fusion and radiation experiments, or explode, in the case of materials experiments.

Z's work helps scientists understand how nuclear weapons age. That perennial concern, the (sensible) obsession of the nuclear complex, moti-vates much of its infrastructure. "We built them, and they've been sitting on a shelf," Sinars says. "How long can they sit on the shelf?"

Every year, Sinars's team does that discharge around 150 times. Each shot requires hours of cleanup, workers scrubbing away black beryllium soot in biohazard bunny suits, because the toxic substance can cause fatal lung disease and cancer.

Z deals with other toxic substances too.

"The thing we're most famous for from an applied science point of view is a highly radioactive, very nasty material," Sinars says.

Plutonium, of course.

"We do not want to be spewing plutonium all over the facility," he says. "You and I would not be walking around here."

So fractions of a second after shooting a pulse of power at plutonium, tiny explosives laid in a ring around the target grind a containment sys-tem closed, trapping the actinide.

To study the aging of weapons, the team takes samples of plutonium from actual weapons and compares their behavior in Z to fresh, youthful batches of the element.

So, how do they compare?

"I can't tell you the answer to that question," Sinars says. (Likely not because he doesn't know but because it's secret.)

Sinars really wants Z to reveal when a given weapon or pit will need replacement. That has implications, he says, for how many plutonium pits NNSA actually needs to make each year—assuming, of course, that aging is the only concern—and if they can get by with fewer.

Sinars is a cautious Type II middle-grounder. "I really do believe that nuclear weapons have, in many ways, made the world safer than it would be without them," he says.

He cites the plot of worldwide deaths in wars, as one must.

"Of course," he adds, "one war would be all it would take to change that equation."


The idea of that one war led to Z's construction in the first place. The lab's earliest interest was in what Sinars calls "hostile radiation effects." Like NIF, aside from just understanding natural aging, NNSA wants to understand whether and how well one weapon could survive the detonation of another nearby. Since the conditions inside Z are close to those of an exploding bomb, you can put bomb materials inside to see how well they survive the effects.

Work like that accounts for much of Z's to-do list. Around 8 percent of Z's time has gone to collaborations with university scientists over the past decade. "What we're able to do here with our university partners is actually create some of those conditions that are hypothesized to exist on the other end of the telescope," Sinars says. Z produces X-rays that can help astronomers understand the plasma sheaths around dense, old stars called *white dwarfs*. The centers of Jupiter, Saturn, and the sun all experience the big-time magnetic conditions Z creates, and its data can show whether digital simulations of those planets behave the way reality does. Though complicated, after all, operating a giant earthbound machine is much simpler than trekking to the core of Jupiter to see if your theories are correct.

But those telescopic conditions inside are also, says Sinars, "very relevant to the center of a nuclear weapon."

Z mostly exists for those more terrestrial studies, and even the basic-science collaborations aren't *pure* in nature. Scientists have studied, for instance, how radiation travels through iron, a quality called its *opacity*. Iron's opacity reveals how different layers of the sun work, and the results of research into it can appear in regular academic journals. There, in public and in peer review, the scientific community can critique the work and so improve it.

National security scientists at the lab can then take those improved results about iron and do their own classified thing for other substances.

Those results don't go into journals. They're another iceberg tip.

Sinars and his team spent a decade on iron's iceberg tip, in fact, and their results didn't always agree with the models astrophysicists used or the lab experiments they did. Scientists, in the university and at Los Alamos, where the digital models were made, would push back. "We don't believe the data," they'd say.

In response, Sinars's team spent a decade making more measurements, reproving the inconsistency. "We could be wrong, but we haven't figured out how we're wrong if we are wrong," he says.

The cries of "wrong!" have grown quieter, but the dissenting scientists now want to measure iron's opacity at NIF, adding its data to Z's measurements, a project that is in process. "So that's where the story is right now," he says.

It's the same story, different book, that Mark Paris and Christopher Fryer told: the models scientists rely on may not be quite right.

*I checked it out*, Sinars could be saying, as Paris did about his simulations. *It doesn't work.*


By March 2022, Paris has moved to a new spot on the Los Alamos campus, a windowless office from which he can do the classified work he's begun to focus on, trying to understand the data and software that illuminate and replicate weapons physics. "I have to pore over it for hours before I get it," he says. "Because I'm stupid."

The new office is a major downgrade from his previous one, but the outside architecture matches the inner workings in a way. The roof is saw-toothed in profile, and the outside of the building is encased in conduits, wires, and pipes—postapocalyptic. "I get scared when I approach," Paris jokes.

He doesn't miss having a window because he never looked out of it when he did. "I go into this hole," he says. "If I want to interact with the outside world, I can. If not, no one knows. No one's knocking on the door. It's great."

Inside that cloistered room, Paris has ingested what he deems "fabulous amounts" of information about "the weapons side of the physics project." While he's long done a bit of that work, he largely focused on the basic-science side. But he sees the lab's structure changing. Whereas basic science used to inform weapons science, things seem, to him, to be transitioning: Paris feels the lab is less interested in deep knowledge of atoms and their constituents and keener about simply engineering and producing objects, like plutonium pits. "There seems to be a perception that there is a weapon science that can be, let's say, 'isolated,'" he says. "Not the greatest word, but 'independent of advances in other fields.'" By this, he means that the lab has separated fundamental science from mission science more directly, studying the latter more intensely without as much attention to or incorporation of the former.

"Which would be a singular event, I think, in the history of science," he adds.

The switch in scientific philosophy, he says, is causing people like him—who have worked on theories of the physical world that apply, sure, to weapons but also to atoms forming after the big bang—to reassess their standing at the lab and perhaps to realign their duties with the lab's priorities. "I was happy doing what I was doing," he would say later. But lab management told him he could only charge so much of his worktime to just-science cost codes. He needed to go find money to cover some of his salary elsewhere. "And I found a lot of money," he continued—where the money always is: in weapons.

Although he's had months to, Paris hasn't much considered the issue that previously seemed to occupy his entire mind: how Ernest Rutherford's scattering equation might be just an approximation ("I've had about four minutes to think about that problem," he says). He's been focused more broadly on trying to infiltrate the "code circle"—the group of people who work on writing the nuclear simulation software. Its ranks can be pretty independent of those who, like Paris, work on the physical theory that underlies their software. The first step in this infiltration is learning to run the codes. The second is opening them up to see what they're actually doing.

But the deeper Paris goes, the more difficult it seems to predict nuclear weapons' states of being and behavior. He also thinks that many of his colleagues don't seek that sort of precision, because the possibility of being able to write and run such complicated code is too far in the future. Thirty years from now, he thinks, with the promise of quantum computers, they might be able to do it.

But still, no matter how precise and complicated, the models will never—could never—be true representations of full reality. The models continue to approach, but never reach, the truth.

How, then, have humans been able to make these weapons that clearly do work—just ask Japan or the Nevada desert—if scientists can't fully, really, truly explain how?

Maybe the same way you can create a wheel without knowing what pi is. Or start a fire without being able to write out the equation for combustion.

Paris sees, in this seeming contradiction, a difference in worldviews. The engineering worldview seeks to create devices that work and then use them. The physics worldview seeks to know what ticks at the hearts of those devices. That we can make bombs but not fully understand them illustrates, to Paris, the gulf between the engineering view of the world—it works—and the basic-science view—here's how.

Maybe both, he concedes, are valid ways of understanding. "Some people see the world through music," Paris says. "Some people see the world through oil painting." Some people get meaning from and communicate using math. "I don't have this thing where I think science is the best way to communicate. I just—" He pauses, gazes upward. "That's the only thing I've ever had any lick of decency at. And I'm not winning any prizes, right? I just like it."

And while musicians and painters and engineers may find an understanding that feels satisfying to them, physicists like Paris are essentially doomed to never grasp their goal. They can grow closer and closer to truly comprehending what makes the universe and its weapons tick. But they may never reach it, the horizon so far away it just seems to recede.

# CHAPTER ELEVEN

Nuclear secrecy is a special kind of secrecy, because the atomic bomb is a special
kind of bomb.

—Alex Wellerstein, *Restricted*
*Data: The Nuclear Secrecy Blog*

HOW THE UNIVERSE ACTUALLY RUNS IS A SECRET THE COSMOS ITSELF keeps. But there are also secrets the Department of Energy (DOE) and the National Nuclear Security Administration (NNSA) keep from the public. Some (many) of those are justified: you just can't have everyone running around knowing how your nukes work. But others, at least in the minds of people like Los Alamos National Laboratory (LANL) physicist Tess Light, could use a little sunshine. Light is often frustrated by the oversecrecy with which the nuclear complex treats its information. Even information that doesn't actually relate to weapons.

The sensors she uses to pick up nuclear explosions from space don't just catch explosions. They also reveal how Earth's ionosphere, the layer of the atmosphere that's made of charged particles and affects radio communication and navigation, works in detail, as well as the ins and outs of lightning strikes, whose bursts show up on her instruments.

In the past, she's written journal articles about lightning—in fact, nearly all of her published work, since she left the field of astronomy, has been about the phenomenon, her own personal tip of the iceberg. She's also been able to extract "distilled data" from her satellite sensors about the ionosphere and make it public. Lightning data have little relevance to national security—and, in fact, the lightning patterns stamped with classification ink could help with determining where weather is about to grow more dangerous. About twenty minutes before a storm becomes tornadic, the rate of lightning production increases suddenly. Hurricanes show the same *jump signature* in their lightning patterns before they intensify. If there's not a radar nearby that can take stock of those storm cells, instruments in space "can still see the lightning-rate increases and know what's happening," she says. Putting that data out would be of value to the taxpayer, who pays her salary and for the sensors that gather that information in the first place.

But the raw information that flows down from on high is classified, likely because it reveals the capabilities of the sensors that collect it. "Sources and methods," as they say in classification circles.

New data from the sensors on the GPS satellites, including about ultrasecret lightning, rest within that dark sphere.

"It's born classified," she says, almost laughing. "But it's lightning."

THE DESIRE TO HIDE ATOMIC INFORMATION HAS LONG FASCINATED HISTORIAN Alex Wellerstein. When he first started college at Berkeley, he was surprised to learn that the University of California had run both the Los Alamos and Lawrence Livermore labs. "It was a little out of sync with my understanding of Berkeley at the time," he says. His understanding being that the school was interested more in antinuclear protests than in bomb maintenance.

But reality, as is its wont, was more complicated than his initial estimation. Down Telegraph Avenue, after all, banners touted Berkeley's Nobel Prize winners, who'd worked on plutonium and uranium enrichment. Their fabric swirled in a haze of left-wing, peace-loving pot smoke. "I thought 'What a strange world that these two things are sort of in the same space,'" he says. It was a kind of superposition—a place with two potential faces.

Wellerstein has gone on, since those college days, to become academically obsessed with nuclear weapons and secrecy. On his website, appropriately called *Restricted Data: The Nuclear Secrecy Blog*, you can click on a page titled "NukeMap" and detonate a fictional bomb—its yield your choice—to see how it would affect a city that is also your choice.

"Drag the marker to wherever you'd like to target," the site invites.

On the site, Wellerstein has, as of mid-2023, written nearly three hundred blog posts about nuclear weapons, beginning with one that explains the blog's purpose. "Nuclear secrecy is a special kind of secrecy, because the atomic bomb is a special kind of bomb," it says. "Just as the atomic bomb has been treated as something above and beyond any other category of warfare, so has its secrecy. . . . When the bomb was thrust upon the consciousness of the world, again and again it was emphasized that it was built by science and by secrecy."

But the connection between the two isn't firm or static, a fact that Wellerstein went on to research in great detail over the next decade, that work resulting in a book titled *Restricted Data: The History of Nuclear Secrecy in the United States.*


SECRECY HAS ALWAYS BEEN PART OF THE NUCLEAR STORY. AND PUBLIC perception of their twining is so pervasive, says Wellerstein, that "if you talk to people about nuclear weapons, you always get somebody saying, 'Uh-oh, are they listening to us?'"

The nuclear quietness embodies what Wellerstein calls "totemic secrecy," the sort that binds people into a clan. "You can see it when you talk to people with clearances," he says. "They begin to believe that only the people with the clearances know anything anyway." Having the right badge is like being in a fraternity, or a secret society, or even being let in on a friend's closely held history: you're special because you're part of something other people aren't.

But hushing science has never really been a reliable way to keep weapons from proliferating or improving in other countries. After all, an atom is an atom, fission and fusion are splitting and combining, and the fundamental laws governing nuclear interactions are relatively easy to grasp, especially almost eighty years after the first atomic weapon was devised.

"If it is based on science, you can have scientists reinvent all of that," says Wellerstein. "At best you're going to slow people down a little bit." Put the brakes on, not stop them.

CHRISTOPHER FRYER AGREES AND THUS SEES MUCH OF HIS EMPLOYER'S secrecy as futile. "The goal is not to keep everything from everybody else," he says. "The goal is to be faster than them." In Fryer's view, then, sharing (some of) Los Alamos's knowledge and stealing public advances for the benefit of the DOE and NNSA lifts both boats more quickly—which makes the United States faster than other countries. That, then, increases security, through openness rather than secrecy.

Keeping LANL things behind LANL doors might have worked a long time ago, when national security scientists were the only ones really developing detailed, multiphysics codes—software that incorporates, say, electromagnetism, temperature, and hydrodynamics all at once, not just one aspect at a time. As Fryer puts it, back then his ilk was "the only horse in the barn." But with the boom in modern computing and increasingly complicated software, new horses—academics, industrialists—are edging in.

That's good for Fryer: stockpile stewards like him must solve more complex problems than ever, and the more brains boosting the field, the better. But sharing has required a mind-set shift for LANL.

The shift is happening under the lab's feet, encouraged by some of the very people who once worked within the secrecy structure. Fryer recalls a recent American Physical Society meeting that included former Los Alamos bigwigs. If you have a lot of gold, they pointed out, and you build walls to protect it, great: the gold's value will continue to rise. It's gold, after all.

"Science works exactly the opposite," Fryer recalls former LANL director Sig Hecker saying. If you build a wall around scientific information, its value goes down. And the wall doesn't actually stop people from discovering or reproducing science. Especially, science that is, at this point, often decades old.

Putting that view into practice, though, hasn't been easy. Fryer has been advocating for making more of LANL's multiphysics codes—or at least parts of them—available to the broader scientific community. "There's been a lot of fights," he says. But he has gotten a few software sets out there. Their names—SuperNu, Semi-Analytic, CCSN—sound inscrutable and innocuous, like all good products of a secret city. Lately he's been able to get them cleared for publication faster than he has in the past—but not quickly or frequently enough for his taste.

"You will have me still fighting this battle," he says, "probably until I retire."

THE NUCLEAR WORLD'S BAKED-IN OPACITY PRESENTS PROBLEMS BEYOND "slower progress," Wellerstein points out: engaging in excessive classification risks running over democratic principles—like the idea that good governance depends on accountability and oversight. "It can lead to a very dangerous situation where people are completely ignorant of what's going on," he says, "or that, behind the scenes, you have incredibly dangerous, risky behaviors that most Americans wouldn't be happy with."

Besides, secrecy has sometimes resulted in more danger, not less. For instance, during World War II, physicist Leo Szilard led a campaign to withhold sensitive nuclear science from publication, lest it give enemies ideas about nuclear weapons and tip them off to the United States' Manhattan Project. Scientists essentially stopped putting out papers on topics relevant to nuclear weapons. But the sudden disappearance of certain subfields and authors from the scientific literature actually tipped foreigners off to the bomb project's existence.

During that project, the media was complicit in the cover-up, largely adhering to a voluntary gag order—which perhaps makes it more of a compelling gag suggestion—on certain scientific topics, like fission. One journalist, famed Pulitzer Prize winner William Laurence of the *New York Times*, actually got access to Manhattan Project scientists as the effort was underway, wrote stories that were held until the project became public, watched the Trinity test in person, wrote the cover-up press release about it, wrote the press release to come later about the real story, and composed the first draft of Harry Truman's speech after the bomb dropped. He was a propagandist posing as a journalist and a journalist posing as a propagandist.

The Atomic Energy Act created the Atomic Energy Commission (AEC), whose functions largely became the purview of the Department of Energy, dropping control of nuclear weapons into civilian hands. The legislation, though, also codified atomic confidentiality. Everything about nuclear weapons was "born classified," secret from the moment of its creation, regardless of who created it. Regardless of whether it was lightning. The classified science that has resulted from this clampdown is ominously called *black literature.*

The AEC and the DOE took secrecy very seriously. In 1950, for instance, the AEC compelled *Scientific American* to cease a print run of magazines and burn three thousand completed copies because the issue contained an article about hydrogen bombs. In 1988, Los Alamos hunted down ten thousand copies of its lab newsletter because officials were concerned that its summary of the director's yearly "State of the Laboratory" speech contained a hint of classified information. Officials searched offices all night, removed the newsletter from its distribution boxes, and asked workers to give back the pages in their possession. Soon enough, they realized the newsletter didn't have anything supersecret inside, and the issue was rereleased.

Today, the DOE classifies much, but not all, nuclear information. It is so-called restricted data, after which Wellerstein's work is named.

But the classification system presents potential problems for scientists, both professionally and philosophically. Science is, in theory, supposed to be open: transparency is one of its fundamental tenets. Without the ability to see in detail what other researchers have done and confirm or refute it, the scientific community (and the rest of the public) can't assess its viability or provide detailed criticism. As importantly, though, if information stays in locked boxes, knowledge cannot build on itself as efficiently. Progress can stagnate. Efforts might be duplicated. Researchers might walk down blind alleys many times, not knowing anyone else has trod them before. And collaboration—except among those with the right-color badge—suffocates.

Within the nuclear establishment, scientists sometimes toil in obscurity because no one else knows about their work, which doesn't appear in traditional scientific journals (not a tragedy, but not necessarily professionally pleasant). The typical "publish or perish" mentality of academic culture is, in "mission work," moot—and mutates into something more like "publish at your peril," because making the information public is often illegal.

Still, scientific openness is more of a virtue in theory than in practice, and criticism of nuclear secrecy is, at least in some senses, hypocritical and in bad faith: even outside classified settings, scientists keep their work hushed so that others don't scoop them. Companies doing research often deem it proprietary. Academic journals ban authors from speaking to the press before publication time, and they place embargoes on articles so that journalists can't write about them before they go live. "Historians, including myself, are always skeptical that science requires an openness to thrive and survive," says Wellerstein. Skeptical, because openness is far from the norm.

Yet, it may be, at least in part, that this is a nice, clean story scientists tell themselves. Wellerstein is familiar with these kinds of entrenched narratives. There's the one about how science and technology are the best things to ever happen to humans. No more outhouses! All human knowledge in electronics that fit in your pocket! Rockets to Mars! Vaccines! Ergonomic keyboards! Debates about black hole singularities!

But science and technology also created weapons that could put an end to science and technology—and the humans who devised both—forever.

CLASSIFICATION ALSO SPURS FALSE NARRATIVES OF ANOTHER SORT. "SEcrecy itself is a major generator of conspiracy theories," says Wellerstein. Because of the mystery around Manhattan-era Los Alamos, for example, people—who were not blind and could see that something was going on atop the mesa—bestowed many imagined identities onto it. Maybe the mesa was home to a poison gas factory, a spaceship maker, a Russian submarine base (inexplicably far from the sea), a pregnancy base for Women's Army Corps members, a whiskey mill, or a camp for rogue Republicans, all possibilities tossed around, according to Ferenc Morton Szasz in *The Day the Sun Rose Twice*.

In the absence of evidence, it's only human to make up stories.

In more recent memory, conspiracies about the DOE's doings show up in the television show *Stranger Things*. Set in a nostalgic version of Cold War Indiana, the drama centers around the fictional Hawkins National Laboratory, a DOE facility somewhat hidden within a menacing forest. The lab can send people to a previously inaccessible dimension called The Upside Down, where monsters and menaces threaten humanity.

Other dimensions, of course, aren't exactly the DOE's purview. But it's easy, and probably correct, to say that if extradimensional research were going on, it would be classified.

The show's strange conceit mostly works, though, because many people don't know the DOE exists, let alone what it does. And on the off chance that they do know, they also know that the institution rarely shares that with the public. So monsters might not be too far off.

Besides, many people also consider all governmental secret keeping to be of the same species, a fact most evident in the QAnon phenomenon. "Q Clearance Patriot," now known simply as "Q," first posted on the message board 4chan in 2017. Q spewed insider-based revelations and gained a large following because he spilled (false) secrets about Donald Trump–hating elites who ran the world and also a pedophile ring.

But Q's access to state secrets doesn't make much sense: "Q" isn't a military or intelligence clearance at all. It's a specifically nuclear, very civilian one, part of the Department of Energy's own specific clearance system, which is separate from the Department of Defense and intelligence community. Q should have been talking about warhead primaries, fusion ignition, and neutron generators—not political power structures. The power of the phrase "Q clearance," though, is that it sounds mysterious—as if the *most* secret of secrets reside with those who have that level of access.

One nuclear-adjacent conspiracy theory about the elites, though, actually does have a firm grounding in reality: the New World Order. The idea takes many forms, but the general premise is that certain elites aim to create a world with no nations, a one-state planet governed (usually in totalitarian fashion) by the powerful people who brought it into being. In nuclear weapons' early days, some of the atomic scientists *did* in fact think that was the way the world could soon lean. A unified federation—perhaps run by scientists—was the only sensible way to govern a planet that had planet-destroying weapons.

Like the New World Order, false conspiracy theories about facilities called deep underground military bases (DUMBs) have roots in nuclear reality. A network of DUMBs, the story goes, crisscrosses the continent, connected by tunnels. Inside them, the military does research on aliens, mind control, and so forth.

One plumbing of the idea goes back to the 1995 book *Underground Bases and Tunnels: What Is the Government Trying to Hide?* by Richard Sauder. A real device, though, has been instrumental to its continued existence. Called the nuclear subterrene, it truly was invented at Los Alamos, in the bright mid-century era when scientists thought nuclear might be good for, you know, everything. The subterrene aimed to bore tunnels for construction projects by melting the earth with nuclear energy. It was never used or built in a nonprototypical way.

Or was it? Are They hiding it from you? If They wanted to keep their bases secret, couldn't They?

"Once you start going down a path on which you're really committed to not taking any form of information as authoritative—even if that's for really good reasons, even if you've been burned in the past—you undermine the ability to think about anything," says Wellerstein.

BUT, OF COURSE, PEOPLE HAVE GOOD REASONS TO MISTRUST THE DOE. The historical nuclear complex doesn't have a good track record of treating humans well. During experiments at the Nevada Test Site, for example, human experimentation took place so that doctors could evaluate the bombs' physical and psychological effects. At Los Alamos, a tissue analysis group once did postmortem testing on workers who donated their bodies to see how their work might have affected them.

The human-harming sin is original, beginning with the Trinity test. The experimenters didn't know much about radiation's effects back then—after all, no one had ever created such significant quantities of it, and not long before women were licking radium paintbrushes just so customers could have glowing jewelry. Still, the scientists knew their bomb wasn't entirely safe: that's why they had evacuation plans in place and wanted low wind to minimize the fallout's spread. In fact, of course, the bomb's uniquely bad dangers were the reason they were building it in the first place.

But, it turns out, they hadn't taken a correct census of people who lived nearby and were surprised to learn that a region they took to calling "Hot Canyon"—for its high radiation readings after Trinity—was actually home to citizens.

Populated, too, was Chupadera Mesa, the valley's prime grazing area. One rancher near the test, Ted Coker, stood outside as the Trinity fallout came down, telling the scientists who came by later that "it smelled funny," according to Szasz in *The Day the Sun Rose Twice*. His neighbor's daughter was playing outside. "At a crossroads store, an old man remarked, 'You boys must have been up to something this morning. The sun came up in the west and went on down again,'" writes Szasz.

Decades later, the federal government instituted the Radiation Exposure Compensation Act, giving medically minute amounts of money to people who could prove they lived "downwind" of Nevada nuclear tests, worked in the uranium industry, or participated in the on-site testing and contracted an illness, like cancer, linked to radiation. No such compensation has ever been available to those who lived near Trinity, although there is a current attempt to expand the coverage to include these downwinders of New Mexico.

ON WELLERSTEIN'S "NUKEMAP," YOU CAN CHOOSE—FROM A DROP-down menu—to simulate how a bomb the size of the Trinity device would impact a given metro area. If you blew up a Gadget over the middle of Denver, for instance, nearly fifty thousand people would die. Another fifty-nine thousand people would be injured. The map shows the blast radius in rainbow-colored contours, revealing a city spurned. Switch your choice to a more modern weapon option, though, and nearly 188,000 people could be snuffed out.

Today, the consequences of DOE's past work mean that it is responsible for 85 percent of the federal government's environmental liability. That involves "remediating contaminated soil and groundwater, deactivating and decommissioning contaminated buildings, and building facilities to treat millions of gallons of radioactive waste," according to the Government Accountability Office. The estimated future cost for that penance was $512 billion as of 2020.

The DOE's Office of Environmental Management is on the hook for the most substantial cleanup effort in the world, tasked with rehabilitating 107 places spanning the continent, an acreage the same as Rhode Island and Delaware, together. While ninety-two spots have been cleaned up, fifteen active ones remain. The nuclear complex is an ouroboros in that way, the tail of contamination being eaten by the newly born cleanup, just as the tail of aging is ingested by the modernization mouth.

But focusing only on the ills and secrets of the past, and the intersections of the two, isn't helpful either. "Nuclear weapons are not just a thing from the Cold War," says Matt Korda, senior research associate at the Federation of American Scientists, an organization that concerned Manhattan Project scientists founded in 1945 in part to shape the nuclear future and minimize risk.

Today, Korda is trying to bring more safety to the topic—by decreasing secrecy. He's the manager of the Nuclear Information Project, which tracks global arsenals and nuclear developments across the world—things that normally remain pretty opaque—using only open-source, unclassified data. The data get published in a document called the *Nuclear Notebook* every couple months. "We try to pull back that curtain of secrecy a little bit," he says.

In some cases, like specific diagrams for how to build your own basement bomb, even Korda agrees that secrecy is legit and needed. But often, he continues, secrecy is maintained "specifically with purpose of limiting public debate about nuclear weapons."

That doesn't just limit citizens' engagement: it means that what's going on within a country is also opaque to other countries—making it hard for them to read each other's capabilities and motives.

The notebook, Korda says, "offers governments a way to talk to each other without revealing anything about their nuclear arsenals that they would be disinclined to reveal." It does the talking for them.

Dealing with this potentially world-ending information all day could be taxing, but Korda has a healthy coping strategy for thinking about this particular disaster. "Completely just dissociate," he says.


DISASTER, THOUGH, CAN LURK ANYWHERE ON THE SMALL SCALE. ON AN unseasonably warm March 2022 day, outside LANL's Center for Nonlinear Studies (CNLS), sits a picnic table nearly folded in half. A tree fell during a recent windstorm, almost splitting it. A foreseen—predicted—weather event nevertheless resulted in specific, unforeseen consequences.

"We're lucky it fell that direction," notes Fryer, grabbing a seat at a different table. If it had gone differently, the tree might have ended up damaging a building or harming people. The difference between those two outcomes can seem like chance, but it, too, was predictable—at least, if you knew the initial conditions perfectly.

Fryer glances over at the CNLS building, where he goes often to get coffee and eavesdrop on the break room conversations of the postdocs who work there. These researchers study all kinds of systems that scientists call nonlinear—those whose outcomes change radically based on just slightly different initial circumstances. Weather is one example: a small change in atmospheric conditions can result in wildly different snowfall. Those systems are hard to predict because you have to know exactly how things were set up ahead of time.

The scientists who work here come from academia or the computing industry, and although their research topics are nonnuclear—turbulence, fluid flows, the spread of Covid—the computational tricks they use to understand and presage are similar. And, of course, the techniques apply well to nuclear weapons, living their nonlinear lives in the stockpile. "We take my problems," says Fryer, referring to stockpile stewardship work, "and their techniques and put them together."

LANL founded CNLS for some of the same reasons Light and Josh Carmichael undertook their project: working in stovepipes—keeping information within narrow channels, hindering its flow in an attempt to keep it contained—doesn't work anymore. Fryer is thus passionate about keeping the lab connected to computational developments in the outside world, where such work has caught up with (or, in some cases, surpassed) that of Los Alamos. "What we are doing was state-of-the-art thirty years ago," he says.

Now more computing specialists don't work for the DOE than do. And giant, powerful companies are investing in such work. The calculus has changed. "We still have managers that don't realize it has flipped," Fryer says, frustrated. "Some people would argue we've already fallen quite a bit behind."

And that, he believes, is because the lab isolated itself—in a desire for security and secrecy and out of a kind of hubris that its work was and always would be the best. Because of that, some LANL scientists haven't kept up with scientific literature that comes from outside the lab and also haven't published as much of their own work as they could have, to get the feedback and criticism of those who, it turns out, might know better.

To illustrate this, Fryer brings up a scientist named Josh Dolence, whom he collaborated with on a project that used hydrocode—software that simulates how fluids flow. Dolence informed him that the lab's hydrocode was outdated—better versions existed, for free, in the open, and LANL scientists didn't even know about them.

Some managers do know that scientists need to keep an eye on the outside world and steal its techniques. But when deadlines loom, insularity can overtake them. "Push comes to shove, and they're trying to meet milestones, and that goes out the window," Fryer says of the motivation to stay connected to developments outside the lab.

The issue, though, isn't just keeping up with the computational Joneses—it's that Fryer and others like him actually need to tackle the thorniness within their existing codes. "Our stockpile is aging. We're going to have to rebuild things," he says. "That is going to require new understanding."

But new understandings require will. "You get old and you don't want to change," Fryer says. He cites his own desire not to learn quantum computing and his resolve, nevertheless, to do so anyway. "There's a reluctance to evolve," he says, "and for some areas we just haven't."

Some crotchety scientists say they never concerned themselves with such gritty problems when *they* were your age.

"Yeah, you didn't worry about that twenty years ago," says Fryer. "Because you couldn't."

# CHAPTER TWELVE

One of the significant outcomes of the
Atomic Bomb Project and particularly
the Los Alamos branch was the bringing
together into a smoothly functioning team
of the long-hairs and the short-hairs, who
in normal peacetime used to growl at each
other from a safe distance.

—William Laurence, *Dawn over Zero*

GIVEN ITS OBSESSION WITH SECRECY AND PROCEDURE, PERHAPS IT'S NO surprise that the events of 2004 sent the Los Alamos National Laboratory (LANL) into panic mode. Two things had happened: One, computer disks containing classified information had disappeared. Two, a laser had injured a student's eye. Neither of these events was good for a facility that needed to look like a tight ship.

In the wake of these apparent breaches, lab director G. Pete Nanos paused all work at the facility, calling lax lab employees "cowboys and buttheads" at a subsequent staff meeting.

"This willful flouting of the rules must stop, and I don't care how many people I have to fire to make it stop," Nanos said in an all-hands meeting.

"If you think the rules are silly, if you think compliance is a joke, please resign now and save me the trouble. If I have to restart the Laboratory with 10 people, I will."

As his immoderate tone implies, Nanos was responding to more than just the two incidents themselves. He was also reacting to politicians' existing concerns about Los Alamos's safety culture. "You may already have seen media accounts of what individuals are saying about the laboratory and these recent events," Nanos said in a memorandum following the meeting. "Perhaps this outside view will help you understand just how serious this situation has become." Some believed the situation stemmed from the lab's academic roots: LANL has always been full of "longhairs," the loose civilian scientists, as opposed to disciplined "shorthairs," or the military types.

The disks turned out not to exist. A similar situation had occurred four years prior, deemed serious enough that the feds came on-site, attempting to get the classified material back under control, even though the disks had appeared behind a copy machine a few days after their disappearance. "FBI agents descended on Los Alamos, administering polygraphs to weapons scientists, commandeering their offices, and, in some cases, dragging them from their beds in the middle of the night and driving them two hours to Albuquerque for interrogations," wrote anthropologist Hugh Gusterson of the official response.

Given that the 2004 security breach turned out to be seemingly nonmalicious, not everyone saw LANL's problems as direly as Nanos did. "The safety record of LANL in the past decade has not been significantly worse than that of Livermore or Sandia and is actually better, especially with regard to serious—as opposed to minor—safety incidents," wrote Los Alamos physicist Brad Holian in a letter to *Physics Today*.

But Los Alamos nevertheless aroused, and sometimes continues to arouse, more side-eye and suspicion than the other labs: it is the most visible part of the nuclear weapons complex, having played the central role in designing and building the only bombs used on humans. "The lab has been the poster child of ban-the-bomb activism from the political Left," Holian continued, "and from the lab's inception there have been calls to close it or at least terminate its nuclear weapons activity."

The Right, meanwhile, has decried LANL's longhair culture. "Why is there such a thing as 'academic freedom' *at all* in a bomb factory?" Holian asked. "Why are *any* foreign nationals anywhere near the security fence?"

Regardless of what employees like Holian thought of higher-ups' assessments, after the incidents they had to go to training (or, really, retraining) sessions and add more security features to their work. Some employees were suspended, and many lab operations were halted for months.

Unsurprisingly, morale at the lab was low. Employees stuck for-sale signs in Nanos's yard, Gusterson details, and put a profane bumper sticker on his car. He, in turn, built a safe room, scared of disgruntled staff.

No one became violent, but in late 2004, a new blog did appear on the internet: *LANL: The Real Story*. It was hosted by Los Alamos computer scientist Doug Roberts.

"I open this blog with the story of how the shutdown affected my group," wrote Roberts. When Nanos paused work at lab, he continued, Roberts's team lost a three-year contract and the goodwill of customers. The group's deputy leader retired, and then six team leaders left. "Should this be an issue of concern?" Roberts wrote. "You be the judge."

Roberts didn't intend the blog to be a place for just him to air his personal grievances: he wanted other people to air theirs too—anonymously. And they obliged, sending him written thoughts to post. Sometimes, those came in poetic form:

> *Director Pete Nanos has said,*
> *if you do science you are a butt-head.*
> *So he stopped all work,*
> *that moronic jerk!*
> *Now science at LANL is dead.*

The irreverent blog grew a wide audience: members of Congress began to consume its content, even bringing it up in a hearing. There, the online griping itself led politicians to wonder if the lab, with its whiny employees, was irredeemably broken. "Why do we need Los Alamos?" asked

Representative Bart Stupak of Michigan. (Answer: Los Alamos designed much of and helps maintain the nuclear stockpile.)

That level of responsibility was also why Roberts saw the need for transparency and defended his blogging as good for the lab. "I think any process that can't withstand the scrutiny that an open venue of discussion provides probably has some serious flaws in it somewhere," Roberts told NPR, "and that system needs to adjust itself to the fact that free flow of information is in part what this country is all about."

Regardless of what Congress or lab management thought of it, that free flow continued and grew. By February 2005, a sibling blog focusing on Lawrence Livermore National Laboratory (LLNL) had appeared. "Thank you 'LANL real people' for having the courage to speak out," proclaimed *LLNL: The Real Story*'s first post.

Around this same time, LANL's management was in limbo. For about sixty years, the University of California had run the facility. But Congress had ordered a recompetition for the school's contract, partly in response to the lab's perceived cultural issues. And the ultimate winner would take a very different form: a commercial entity. Its more businesslike nature would alter the operation of the lab—and not, some maintain, for the better.

IT MAY SEEM SURPRISING THAT A NATIONAL LAB IS RUN BY ANYONE BUT the nation itself, but it's true: the National Nuclear Security Administration (NNSA) facilities, while owned by the federal agency, are actually managed and operated by private contractors. Those showrunners have sometimes been a university, sometimes a consortium of universities, and sometimes a consortium of private companies and academic institutions. That middleman way of operating has been in place almost since the beginning of the nuclear complex.

After the Manhattan Project wound down and the Cold War blew in, a new version of federal research came with it—similar to but not exactly like the big-science coalition that had built the bomb. Officials didn't want their research and development (R&D) or their purchasing power tamped down by the normal government bureaucracy, but neither did they want them subject to the whims of capitalism. And so they created a new kind of organization: a federally funded R&D center (FFRDC). FFRDCs would be technically private but would do almost all their work for the federal government.

The first such center came out of the US Air Force in 1947: it was called RAND, a proper noun meaning "R&D." The nuclear labs founded during the Manhattan Project soon operated under this same "government-owned, contractor-operated" (GOCO) model. AT&T started running Sandia; General Electric managed Oak Ridge; the University of California took over Los Alamos. "The original contracts were little more than a handshake agreement," says a *Public Performance & Management Review* article. "Early contracts were based on a dollar-per-year fee and were motivated chiefly as a public service."

Today, national telescopes and other scientific facilities are often also run this way—funded by the likes of the National Science Foundation—with a go-between organization, sometimes a for-profit entity, operating them. The national labs of the Department of Energy (DOE) all have this structure. Their workers mostly aren't civil servants: they are employees of corporate conglomerations—chimeric organizations often involving a large defense contractor and a smattering of smaller companies and universities. These collaborations exist almost exclusively for the purpose of operating federal labs. You probably haven't heard their names, but they garner billions of dollars in contracts per year. Modern NNSA lab operators net many millions of dollars annually in management fees alone. And that's a big change from how things began. As a result, the labs' science isn't incentivized in the same way it used to be, the focus being more on businesslike expediency, which can lead to less innovation, more obstacles to innovation, safety problems—and, of course, more federal tax dollars going into the pockets of the organizations that run the show.

The GOCO system has had three historical stages, according to the aforementioned article: the initial handshake, a fee-based contract whose primary benefits are science and technology advancements, and "a system in which the fees are much increased and provide an incentive in themselves." The article posits that in that last phase, where we are now, FFRDCs have shifted focus from science and technology "to a more traditional management process in which accountability, efficiency, and concern with liability were boosted in importance relative to the science mission." In other words, places like the national labs are being run like regular bureaucratic companies.

When the University of California lost the management contract for Los Alamos, the lab entered that third stage, and a strange entity called Los Alamos National Security (LANS) took over the site. LANS was a for-profit company led by the Bechtel Corporation—a defense and infrastructure contractor that boasted billions in revenue each year. The California school system was one of a few partners.

With that change, at the end of June 2006, Roberts shuttered his site.

The words "Last post" hyperlinked to a version of the military death tune "Taps."

BEING MANAGED BY A COMPANY MADE LANL ITSELF MORE LIKE A company.

When the University of California had run the show, it had reinvested much of its management fee—the money the operator got for doing the job, on top of the lab budget—back into the labs. LANS's management fee, meanwhile, increased from $8 million to $80 million between 2005 and 2010. And it didn't pour dollars back into facilities the same way. Bonuses to LANS leaders and from the government to LANS itself were calculated according to managerial metrics—not based on scientific prowess. Layoffs came.

Neither cowboy nor butthead was onboard with the new structure, and someone did fill Roberts's vacuum, starting a new blog called *LANL: The Corporate Story*. One of the first posts, from December 2006, hit the point home. "The name 'Los Alamos National Laboratory' is no longer appropriate for that facility. Something like 'Los Alamos Corporate Laboratory,' or, in a few years 'Los Alamos Corporate Plutonium Foundry' would be more fitting," it read. "It is becoming clear that the bottom-line profiteering orientation of LANS, LLC is completely incompatible with the quality of science that one would expect from a true national lab."

Blogs continued to spawn blogs, with sites called *LANL: The Rest of the Story* and *Gone Nuclear* cropping up. Lawrence Livermore, soon run by a similar conglomerate corporation, got its own online gossip columns: *LLNL: The Final Story* and *LLNL: The True Story*. The latter's logo showed two hands made of money shaking, suit coats riding up their wrists.

But the angrily typed outcries were not enough to stop the changing culture.

"Time to Declare Victory," announced the title of a January 2010 post on *LANL: The Rest of the Story*. "LANS has won; nearly everybody else lost," it continued. The post then showed fake motivational posters from the website Despair.com: A ship sinking into the ocean and the words "Mistakes: it could be that the purpose of your life is only to serve as a warning to others." An image of one arrow splitting another on the outer edge of a bull's-eye, captioned "Consistency: it's only a virtue if you're not a screwup."

EVEN NONDISGRUNTLED OUTSIDERS AGREED THAT THE CORPORATE managers had altered the labs. A 2013 National Research Council report, for example, noted that under the new company-centric management, the costs of operating Los Alamos and Lawrence Livermore had increased considerably, decreasing the money available for scientific research. "If the incentive fee becomes too high, or the criteria upon which the fee is measured discourage experimental science or innovation," it read, "the scientific enterprise at the laboratories could well deteriorate over time."

Another 2013 National Research Council review found that the for-profit management switch at LANL and LLNL frustrated employees, led to many leaving the organization, and upped the contract costs by $200 million. That increase was partly due to higher management fees, which at Livermore and Los Alamos hovered between 3 and 4 percent of the total budget. That meant LANS was getting $63.4 million, which is about $63.4 million more than a handshake.

But just 30 percent of their money was guaranteed: the rest depended on performance evaluations—most of them unrelated to actual science or engineering and instead tied to corporate efficiency. That de-emphasis on research productivity led to "box checking and transactional compliance," according to the Commission to Review the Effectiveness of the National Energy Laboratories, and to a "breakdown in trust between some of the laboratories and DOE."

One researcher, for instance, needed special multiperson permission to attend a very normal physics conference. Another employee confessed they did not take on ambitious R&D projects because of the extra hoops they'd need to hurdle through.

Not everyone, though, believes this formality is due to the new contractors. And not everyone believes the solution is more oversight: some maintain that the DOE oversees too much already, micromanaging labs that were meant to be privately operated in part so as to be flexible.

Regardless, the transition toward corporate managers is complete at the tri-labs. Since 2018, Los Alamos has been run by Triad National Security—made up of Battelle Memorial Institute (which saw $5 billion in revenue that year), the University of California, and Texas A&M University. Livermore is managed by the Lawrence Livermore National Security, LLC—another unparseable name that today translates to Bechtel, the University of California, NWX Technologies, and Amentum. Sandia's operator, meanwhile, goes by National Technology and Engineering Solutions of Sandia—a subsidiary of Honeywell International, Inc. (you know, the conglomerate that makes your thermostat). The Savannah River Site is run by Savannah River Nuclear Solutions, a partnership comprising Fluor, Newport News Nuclear, and, formerly, Honeywell.

It is perhaps true that the academic longhairs lacked a buttoned-up business or military culture, but the new species of showrunner sometimes lacks more important things, like scientific expertise and motivation. Those gaps could lead to new and different safety problems.

Just ask the Waste Isolation Pilot Plant (WIPP) near Carlsbad, New Mexico. There, transuranic waste—radioactive detritus containing uranium or heavier elements—lives hundreds of feet underground, in salt beds that isolate it from the environment.

Or so the idea goes.

It was Valentine's Day 2014, and all was quiet on the WIPP front.

Then, suddenly, more than two thousand feet below Earth's surface, a drum full of nuclear waste burst open, starting a fire and letting loose radioactive particles. "Plastics and cardboards burned. The repository's underground Continuous Air Monitor alarms went off. Its ventilation system began automatically filtering underground air. Small quantities of radionuclides escaped to the surface through a defective damper on a vent," wrote nuclear anthropologist Vincent Ialenti in a research article for *Social Studies of Science.*

The waste drum had become a dirty bomb, and its eruption and emission lasted more than two hours. "The next morning, 140 people were on-site, standing outside," Ialenti continued. "They were unaware of what had happened the night before."

Some of those at WIPP that day were DOE investigators who were visiting to look into a different incident: a salt truck that had ignited underground.

In a subsequent investigation, the DOE found that the problem was essentially bad chemistry: the drum contained nitrate salt waste, an acid neutralizer, and a substance called sWheat Scoop. sWheat Scoop is kitty litter, meant to soak up liquid and stop chemical reactions from happening between the other ingredients—but it was the wrong kind of kitty litter: organic, made from plant material.

The recipe that workers had been following to fill the drums had been incorrect: instructions for a new repackaging procedure had said to add an organic rather than an inorganic kitty litter to the drums. The organic material then reacted with the waste, releasing heat and building up pressure over the weeks the ingredients were combined together. That reaction was happening inside every drum in that batch. "As a Los Alamos manager joked, the lab accidentally 'made 700 dirty bombs, but only one worked,'" wrote Ialenti.

And work it did: in the end, twenty-two people received detectable, though minor, radiation doses. WIPP closed for almost thirty-five months. The cleanup cost more than $1 billion.

"It's really seen in public perception as the result of a note-taking error," says Ialenti. But pointing to a simple stenographic issue was, in his opinion, a way to make a few heads roll and sweep the problem away.

The real story, Ialenti believed, was more complicated—relating in part to the lab's management structure and culture, which reward speed and efficiency. One employee he interviewed called part of the problem the "puppydog effect": "The very human potential for pats on the back for speeding things up," says Ialenti.

IALENTI TRACED THE PUPPYDOG AND OTHER ISSUES BACK PARTLY TO A wildfire around Los Alamos—its own problem, growing more trenchant as wildfires themselves grow more frequent and larger. In 2000, the Cerro Grande fire burned forty-three thousand acres, some of them on lab property, and destroyed hundreds of homes in the city of Los Alamos. While no special nuclear material (the kind that becomes fission fuel) was lost or destroyed, Cerro Grande made people reasonably fearful of what a future conflagration might mean for LANL's large inventory of old waste. To get the waste away from the site and into permanent storage elsewhere, LANL created a program called "Quick to WIPP." Its goal was, as the name suggests, to move the waste from one spot to the other. Later, the lab, the DOE, and the state of New Mexico also initiated a 2012–2014 program called the "3706 Campaign" to move more waste—3,706 cubic meters of it; thus the name—away from Los Alamos, away from potential fires, to WIPP. According to Ialenti, if the lab's management met tight shipping deadlines, they'd get performance bonuses.

Money, wildfires, political concerns, lack of scientific knowledge, and the need for speed had come together and also been compounded by regulatory gaps and a lack of oversight: it was the perfect set of ingredients to yield a blowup underground. "This is the product of schedule pressures," says Ialenti. "This is contractors trying to get their performance bonus. This is the product of governors trying to appease their constituents and pressure the DOE to get waste out of there in a certain amount of time. This is a product of the DOE trying to show they can get something done."

It was pressure, in other words, all the way down.

NOT LONG AFTER THIS INCIDENT, AND NOT UNRELATEDLY, LANL GOT A new corporate operator: the aforementioned Triad National Security. "NNSA awarded the new contract in the midst of increased concerns around safety at LANL as it undertakes an expansion of its nuclear weapons missions," says a 2022 Government Accountability Office report, referring, in large part, to plutonium pit production. The Triad contract included a clause specifically calling on the company to make the lab less dangerous.

Federal concern about plutonium operations isn't misplaced or without recent history. On-site, pit production will happen in a part of the lab known as Tech Area-55, inside its Plutonium Facility-4 (PF-4) building. This is the nation's premier plutonium-handling site and virtually the only place on campus where Los Alamos employees mess around with that enigmatic substance. Perhaps not surprisingly, it sits in the interior of the lab's span. Tech Area-55 is hard to reach if you're not supposed to be there, and any potential missteps are thus supposed to be easier to contain. Its 233,000 square feet—the size of almost five football fields—are designed to withstand hurricane-plus winds, as well as seismic shaking. For its pit production work, LANL is making infrastructural improvements: upgrading the gloveboxes that workers use to manipulate plutonium, increasing resilience to earthquakes that might be more likely than people initially thought, installing new fire protections, and upgrading power and ventilation systems. The lab is also cutting back on the plutonium normally stored at PF-4, since so much more will be incoming.

But PF-4, where workers will soon(ish) construct a lot of pits, has been the subject of accidents, criticality safety investigations, a nearly four-year related shutdown of parts of the facility, and concern from the Defense Nuclear Facilities Safety Board that the building and its equipment aren't hardened enough against some natural disasters.

Wherever you go, there you are: even with a new contract operator and plans for a better building, the lab hasn't avoided plutonium mishaps. Triad has improved things, the Government Accountability Office noted in a 2022 report. But the occurrence of incidents like the one in winter 2022 suggests that "lessons learned from errors under the prior contractor have not been fully integrated into laboratory operations."

Just ask the six LANL employees who went to work at PF-4 on the morning of January 7, 2022.

PERHAPS WHEN THEY BADGED IN THAT WINTER MORNING, THE EM-ployees were thinking about the possibility of danger. After all, it's always there: plutonium operations are risky, just as rocket launches are, no matter how routine they seem. But perhaps, because routine is routine, the workers were just thinking of doing their jobs, like always. Their job that day—for which they were briefed on arrival—involved repacking legacy radioactive materials, or ones left from previous work. So they put on their personal protective equipment and checked that the equipment they'd be using looked shipshape.

The primary tools they were using, and the ones that ended up causing them problems, are called *gloveboxes*. The devices are essentially highly shielded rectangles, with openings connected to thick gloves. Workers slide their hands into these gloves, as you might into a box at a Halloween party into which the host has put peeled grapes (eyeballs) or spaghetti (intestines). And, thus protected, they can manipulate the material.

Two hours into the job, one employee had finished work and so pulled their hands from the gloves. Right away, an alarm that senses alpha radiation—helium nuclei flying around—went off. A coworker put in a request for the radiological control technician team.

A minute later, the air monitor sent its warning out, notifying the broader operations center and further alarming the workers. The six employees scurried out of the room, which was then "red lit" so no one else would go in. In the room next door, radioactive readings also shot up. That room, and the one on the other side, went red.

Soon radiological control technicians showed up, surveying the employees and the room itself to see who and what was the problem.

Two workers' protective equipment and skin showed contamination. The one who'd had their hands in the affected glovebox had elevated radiation counts coming from the right torso of their coveralls, their Covid mask, and their face—particularly their left jaw. The second employee had a radioactive lanyard, a radioactive face, radioactive hair.

The technicians took nasal swabs from all six employees to see if their bodies had taken any radiation inside.

Three had.

Within ten minutes of the incident, the technicians flew into external decontamination mode—which, on the outside of a person, involves washing with soap and water, physically removing the offending particles. Once the affected person is dried off, you can survey them again for continued evidence of contamination, rinse, repeat. The workers were successfully rid of the offending particles, according to a safety board report.

After that, the workers weren't allowed to enter PF-4 until further notice. All six went on a "special bioassay" program. That's fancy language meaning that they had to submit urine samples to determine whether they'd ingested radioactive material, how much, what kind, and what sort of dose that might yield over the next fifty years. The employee who'd set off the alarm was then given chelation therapy, in which they would take drugs that bind to heavy metals and drag them out of the body through the urine.

Once the air monitors showed normal levels of radioactivity, the response technicians went back into the affected room themselves, seeking the contamination source. Easily enough, they homed in on the high-dosed employee's glovebox and the glove itself. The room was contaminated with americium-241 and plutonium-239.

And perhaps that is part of why, although the medical results from the six workers didn't show contamination so high that a formal accident investigation was required, NNSA initiated one anyway, turning in a report that same spring. The other part of "why" is that this PF-4 event wasn't totally isolated: it followed other glovebox breaches in the previous two years, while Triad was in charge.

The resulting investigation revealed that the ventilation system was set up incorrectly, and the glovebox wasn't isolated from it, so radioactive material snuck through a degraded gasket and into the air. That was, the report said, the "direct cause" of the accident. But just as a typo about kitty litter may have been the direct cause of the WIPP explosion but not necessarily the whole story, an equipment mishap was not the root of the problem.

There wasn't enough management of gloveboxes; there was too much confusion about who was in charge; there were too few protocols; and there was "overreliance on the skill of the worker," according to the report. Plus, officials had known about "questionable seals" in gloveboxes for a decade, the document noted, and hadn't fixed them.

None of that sounds quite cowboy or quite butthead. In fact, it sounds much more middle manager.


On a less radioactive part of LANL's campus, David Clark has an idea for how to make plutonium operations safer—by making pit production more science than factory minded, more understanding than deliverable driven. In his vision, the hoped-for change in safety culture comes not from paperwork but, in a way, from homework. As director of LANL's National Security Education Center, he designed and helps run a summer school for the people who will be working in the plutonium facility: the one-year Nuclear Enterprise Science and Technology academic certificate program with the University of New Mexico.

Much of the training for those working with plutonium is currently what Clark calls "compliance driven." "I can take my test with the open book, and I can pass, and then I can go into the facility, and then I can have an accident," he says. But if technicians learn not just what checklists look like but understand what's actually going on with the actinide element, they might do better work, because they understand the science behind it. "It's part of our vision for changing the culture in our nuclear facilities," he says.

One of the education courses is titled "Actinide Science Fundamentals." "What are the chemical and physical characteristics of the materials?

What's the fundamental behavior of the chemicals that they work with?" Clark says. The curriculum goes through plutonium's life cycle, from the nuclear reactor that creates it, to the metal manipulation that shapes it, to the waste management that deals with its consequences. With that knowledge, workers might be able to recognize problems that crop up. If the plutonium turns purple, they might know that it's gained extra electrons instead of losing them, like it's supposed to.

They also take a class titled "Nuclear Facility Fundamentals," where they might learn about what size particles they could inhale at work and so why they wear a respirator and why it's different from one you might paint your house in. Students also take field trips to LANL's plutonium facilities and hear from existing employees how gloveboxes and other equipment work.

After workers finish six total courses, they earn an academic certificate and are eligible for a raise (although Clark says the lab is still working on making that eligibility a reality), and they're also eligible to go take more classes toward an associate's or bachelor's degree—on the lab's dime. The lab is hoping to help Savannah River launch a similar program.

Clark believes the lab has generally moved in a safer direction since he first came here. There used to be a lot less oversight and fewer regulations. Today, there's more of both. "There are those things that we do that make a lot of sense, because we don't want to hurt our coworkers," he says. "And then there's a lot of additional bureaucratic nonsense, paperwork and stuff that somebody thinks that if this is good—wouldn't it be better if we added another ten pages? And this is where perfect is the enemy of good enough, and we haven't quite learned that lesson."

## Part Three

# NONSTATES AND THE NEXT GENERATION

Bombs lobbed by other countries and America's maintenance of its own weapons are sizeable concerns for the national labs. But so are smaller dirty bombs, medium amounts of radioactive material coursing through the black market, and the large task of figuring out who's up to what where, and what to do about it. Scientists at the National Nuclear Security Administration labs are working on those issues, but also on something a little more HR: how to get new blood into the complex. The two problems aren't always separate. Sometimes, they're fused.

# CHAPTER THIRTEEN

For those who say that building a nuclear weapon is easy, they are very wrong, but those who say that building a crude device is very difficult are even more wrong.

—Harold Agnew, in a hearing before the Senate Committee on Foreign Relations, March 6, 2002

ONE DAY IN THE 1970S, THE MAYOR OF ORLANDO FOUND A NOTE ATtached to his car. It claimed that a fifty-kiloton "nuclear fission device" was located somewhere in the City Beautiful. The unnamed organization writing the letter threatened to detonate the bomb unless certain conditions were met, namely "1 million dollars cash in small bills and safe passage out of the country."

"If you decide to, you will hear from us later, but if you decide no Orlando will go up in a big cloud of smoke," they wrote. "This is no bluff. If you think it is, ask the Atomic Energy Commission [AEC] what happened to the shipments of U-235 that never got to their destinations. We can be out of Orlando, Florida, before it goes off, but all the people of Orlando can't."

The typed letter was unsigned.

After a bit of pressure, the AEC admitted to Orlando the unthinkable: some of its nuclear material was, in fact, missing. The threat, which the city hadn't taken super seriously, began to look more credible.

Soon, another note arrived, this time handwritten and including an address where officials should direct their response. This letter, worryingly, included a sketch of the supposed bomb. An expert said it wasn't totally inaccurate.

Getting nervous, and taking the only real clue they had, officials sussed out the address. It was just an empty house.

The structure may have been unoccupied, but, people revealed, a teenager would sometimes come to cut the grass. Officials raced to track down this boy and tested his handwriting: it matched the extortion attempt.

Finding no fission devices and just a well-informed adolescent, authorities heaved a sigh of relief. Also, they placed the boy under arrest.

This threat turned out to be a hoax from a fourteen-year-old with zero bombs. But according to Cameron Reed in an article for *Physics and Society*, between 1970 and 2013, bad actors attempted nuclear extortion within the United States 350 times. After the initial Orlando event and a few other hoaxes, officials realized they needed a plan and a team to deal with the next, perhaps real, threat. In response, Fred Jessen of Lawrence Livermore National Laboratory set up a project called Warmspot. Using a Ghostbusters-type van, the project aimed to pick up and pinpoint sources of radiation.

Warmspot scaled up into a more formal operation—a team dedicated to emergency response—after another threat descended on the city of Boston. Although it had a different parent organization and a different name at the time of its creation, the operation eventually lived under the National Nuclear Security Administration (NNSA) and was called the Nuclear Emergency Support Team (NEST), a group specializing in finding and neutralizing nuclear threats or responding if those threats aren't found and neutralized in time, providing assistance both at home and overseas. On the twentieth anniversary of one of NEST's units, NNSA administrator Frank Klotz said, "Like the personnel who operate our nuclear deterrent, you pray mightily that you never have to perform the mission for which you've trained."

Today, NEST has split into subgroups, with one of its business ends being the Joint Technical Operations Team (JTOT), whose members deploy to the scenes of events. JTOT, according to a Los Alamos presentation, "will respond to nuclear terrorist threats anytime, anywhere," sniffing out and ideally deactivating danger, in a process NNSA calls "render safe."

Given the skewed ratio of hoaxes to real problems, the response team now also has the Nuclear Assessment Program, which analyzes whether a threat is realistic. To keep a preventative eye on threats that might exist in the near future, the Department of Energy (DOE) has helped gather data from detectors at ports and border crossings, attempting to determine whether the data point to something benign (like a boatload of bananas) or threatening (like a fistful of yellowcake, a form of uranium that can, with further processing, become weapons grade). The agency also deploys a fleet of aircraft—helicopters and planes, together called the Aerial Measuring System—that take radiation measurements on call 365/24/7 and also do flyby or hover checks on high-profile events, like presidential inaugurations and the Super Bowl.

It's a tough job—which you can tell from a JTOT training presentation that provides the team's most important rules.

Number one, "Look out for others . . . but assume no one is looking out for you!"

Also, "Hydrate or die."

For the most part, NEST team members aren't out sniffing down dirty bombs in real life, and in fact no dirty bombs or terrorist nukes have been used anywhere, at any time in history. NEST has responded to hoaxes, to nonradiological devices, to a nuclear-powered satellite that crashed through the atmosphere, and to the Fukushima nuclear power plant disaster, among other things. But none of those is quite like the real thing—what the team was actually created for. And so personnel get much of their experience in simulated scenarios—fictional circumstances they have to physically work through. It's kind of like LARPing—live-action role-playing—for a nuke SWAT team. It's also like a lucid nightmare.

One such exercise series is called "Cobalt Magnet," which NNSA calls a "play" on the phrase "consequence management"—maybe the most bureaucratic play on words ever. Nevertheless, every three years, a Cobalt Magnet exercise happens in a different city, and personnel—from NEST's branches, from other arms of NNSA, and from local organizations—pretend their town is home to an unwelcome, radiological disaster.

In 2022, the chosen city was Austin. Austin had reached out to request that NNSA (and the dozens of agencies involved in that region) run the experiment on their fair metropolis so they could measure their preparedness.

Phil Schneider, NNSA's emergency operations and exercise planner, was in charge of planning that year's exercise on behalf of NNSA. For five days, thirty agencies and four hundred participants played make-believe after being briefed on the following scenario: Violent domestic extremists were threatening the town—famous for country music, barbecue, and rising house prices—with the detonation of a radiological weapon, a dirty bomb. Officials, the briefing continued, had determined that the threat had not just come from a bunch of lawn-mowing teens: the terrorists had the ability and means to get their hands on radioactive material and to use it in a weapon. As for more information about the plotline?

Some of that information, Schneider says, is "a little bit more restricted." He declines to give many details on a public event that had hundreds of participants.

As the mysterious circumstances surrounding Cobalt Magnet unfurled, experts attempted to find the extremists' device. In many of NNSA's LARPs, this search is the point: NEST personnel locate the weapon and render it safe. Game over, in the good way. This time, Schneider says, "we wanted to say, 'We searched, we failed, how do we transition?'" Here, as everywhere, failure is of course a euphemism, eliding the fact that it means a dirty bomb detonated.

Teams from NNSA's Radiological Assistance Program are usually the first NEST personnel on the scene after an event, or a simulated event. NNSA has crafted the tablets they carry around to show the radiological signatures that would be emanating from wherever they are. "It will actually pop up and show them the simulated data," says Schneider. "So they know the true extent of what this device would look like."

Once they get data, which reveal what the illicit radioactive material is, how it is distributed, and where it originated from, they send that information over to a group in California. There, at Lawrence Livermore National Laboratory, the National Atmospheric Release Advisory Center (NARAC) can start making refined models of how that toxicity would travel and what kinds of casualties and consequences Austin could expect.

At NARAC, scientists are tasked with 24/7 monitoring of nuclear, radiological, chemical, biological, and hazardous natural material that gets released into the atmosphere and with predicting how that stuff will flow through a city and around the world. The team, safe in an office building at Livermore, dispenses information to NEST's in-the-field personnel, whether they're in a real scenario or, as is in this case, living out a made-up nightmare.

AT NARAC DURING COBALT MAGNET, SCIENTISTS ARE HEADS-DOWN. Clocks in their operations area show a plethora of time zones, and a *WarGames*-style wall of screens, onto which individual workers can project their findings, looms over the scene. That sounds high-tech, but the whole operation can also work without the internet in extreme emergencies, relying instead on old-school fax lines and landline phone connections. There's even an archetypal red handheld phone.

Lee Glascoe, a slight man with long, curly gray hair, is head of this crew. During Cobalt Magnet 22, in May 2022, his phone constantly buzzes. Expert in screening out data that's irrelevant to the current moment, he never glances at it.

Glascoe is used to these practice runs. NARAC runs exercises or drills multiple times a month, sometimes at small scale and sometimes, as today, at full tilt. "We're simulating the kind of Japan-level stuff that's like, 'Oh, my God, now what do we do?'" he says, referring to Fukushima.

Members of the team are on call 24/7. When disaster strikes, they also call on the resources of the rest of the lab. "We need somebody who has an expertise in cesium," says Glascoe, mimicking a crisis. "Okay, well, this person over here—yeah, wake them up. Bring that person in."

It was Livermore's scientific expertise that led to NARAC's founding in the first place. The organization began, in a less official capacity, during the 1979 Three Mile Island incident, when an American nuclear power reactor partially melted down and later released radioactive gases into the air. As the almost-catastrophic event was in motion, the Jimmy Carter administration reached out to Livermore: Could the lab provide any useful information to teams headed to the affected area? They could, it turns out. In the 1960s and 1970s, the lab had worked on the global circulation model, which showed how air moved around the world. The model could also be refined, harnessed, and directed toward the catastrophe, revealing how contaminants would spread across the world and what their ultimate fate would be.

In the years since, NARAC has had a hand in all the major nuclear disasters you've heard of, like Chernobyl (which Glascoe pronounces like a Russian) and Fukushima. It's also helped with many smaller or near-disasters that you probably haven't heard about. But Fukushima really amped up life for current employees. "I think that was when everybody realized, 'Okay, this is what a real serious response is,'" Glascoe says.

NARAC's software still relies fundamentally on atmospheric transport models, at the local, regional, and global scale, all simulating how materials disperse once they're in the wild, taking into account the complexities of weather, terrain, and point of origin. Sometimes NARAC personnel get geographic coordinates for that starting spot, and sometimes they estimate it using TV camera footage showing "a smoking hole or whatever," says Glascoe.

They start their predictions with meteorological data straight from sources like the National Weather Service, but they can feed it into their own higher-resolution meteorological software to get higher-resolution toxin forecasts. Fifteen minutes after any given alert, all hands are on deck, providing wind predictions and an estimate of where the potential contaminants are right now and where they're headed. Next comes a "smoke plot," giving more detailed information on where the hypothetical material is wafting. Once they have more details about what that material is, the center can do more detailed work.

But NARAC doesn't just need to know where the problematic toxins will be: it also needs to know whom they will affect and how. For that, it has databases of population and terrain showing the locations and shapes of buildings in hundreds of cities and can build grids for other locations quickly, sometimes in less than a day, to understand how human beings may be harmed. The center's software then spits out simulations, movies to help visualize how material will flow through a region. In these movies, substances in a rainbow of colors drift in between buildings. NARAC scientists add and update modeling and data details as the minutes pass and more information becomes available to work with. The resulting data products look more innocuous than their implications. A plume made of pink dots bubbling upward, for example, represents historical data from Chernobyl.

As updates come in, plots coalesce and get updated. Those new forecasts direct measurement teams on the ground and planes and helicopters in the air about where to gather real data, like radiological readings. They pass these back to NARAC, so NARAC can send back another assessment of the hazards. It's cyclical data sharing that reminds one of the swirl of the atmosphere itself.

As a scenario, real or imagined, unfolds, NARAC gets more information about events—from field teams, from the news, from on high. Then the team can sprinkle in, for example, the strength of the dirty bomb's explosion and the particulars of its cesium to provide a prediction that closer approaches reality. The center then produces what Glascoe calls a "final product."

It doesn't look like colorful simulations. "A nice little perspective of the plume blowing down a city street is cool for scientists or cool for a movie or something," Glascoe says. But when people are trying to make decisions, they don't need movies: they need to know whom to evacuate. And so software automatically creates briefings that show air and ground contamination, health risks, and casualty estimates. They don't show the names or faces of the people who live there, who've become a mass of risks and casualties.

NARAC DOESN'T JUST ENTER GO-MODE WHENEVER GLASCOE FEELS like it: the organization has to be "activated," sometimes by NEST and sometimes through a wider government partnership led by the Federal Emergency Management Agency, called the Interagency Modeling and Atmospheric Assessment Center. But he does have enough experience to sense when a task might be headed NARAC's way. "If there's something that is going on, especially with a DOE site, or some international thing, I can usually tell that there's going to be some interest," Glascoe says.

And DOE sites have plenty of things going on. In summer 2021, for instance, wildfires erupted at the Nevada National Security Site. Officials had their eye on a particular area called Buggy, where five nuclear devices had been set off back when explosive testing was a thing. Detonated 150 feet apart and 140 feet underground, they carved a nearly 900-foot-long trench in the ground.

This was one of the peaceful Plowshare tests, conducted to learn about nuclear weapons' adeptness at digging trenches, which left violent craters with some of the highest soil-contamination rates of the whole site.

The fires burned the area up, and officials worried that contamination left from those experiments might be lofted into the air. In a press release from that July, the test site addressed the concerns. "Estimates provided by Lawrence Livermore National Laboratory's National Atmospheric Release Advisory Center indicate potential radiological exposures for on-site personnel are well below the criteria for initiating protective actions," the release read. "There is no risk to health and human safety, and there is no offsite risk to the public."

The public would perhaps have liked to see those official estimates themselves, but the agency's press releases typically link to results from the Community Environmental Monitoring Program, meant for public consumption.

THE LESSONS LEARNED IN COBALT MAGNET, TOO, STAYED BEHIND closed doors, although Schneider says the report filled about eighty pages. That lack of information in the public is symptomatic of nuclear terrorism in general: there's a dearth of data.

Because the threat of nuclear terrorism has never been realized, no one knows how big it actually is. There are sensible ways to think about the risk, though, says Miles Pomper of the James Martin Center for Nonproliferation Studies. He has a qualitative formula for considering how worried people should be about such threats. "The usual way that I think about it is risk is a product of potential consequence and likelihood," he says. "Obviously, with a nuclear weapon, the potential consequence is tremendously high, but the likelihood is quite low."

The chances that any given bad actor will get their hands on full-on nuclear weapons is small. But crafting a dirty bomb is exponentially simpler. Non-bomb-grade emissive substances exist in a lot of places you never think about: hospitals, construction sites, drilling outposts, food factories. "Most countries don't have nuclear material, but almost every country has radioactive material for industrial uses," says Pomper. And it's not always well protected.

Programs exist both to bolster the security at locations that host radioactive material and to decrease the amount of that material in existence in the first place. Why, though, is all this stuff sitting around, risking proliferation, if it doesn't have to be? The truth is, humans built reactors—and blood irradiators and leak-detection systems and food cleaners and cancer treatments—that required dangerous material because they weren't thinking hard enough about the future. They weren't thinking about the potential risks of radioactive material dotting the globe, lodged in everyday devices. But even if they couldn't have ensured a safe future, they could have gotten closer to a truthful prediction of it. "In hindsight, the connection is obvious," say Christopher Landers and Joanie Dix, both of NNSA's Office of Conversion, which helps decommission reactors or transform them into the sort that doesn't require bomb-grade material. "But in the early days after World War II, nuclear technology was viewed as having the potential to solve many of the world's problems."

Now, however, it also presents the potential to create world problems. And some experts, like Nickolas Roth of the Nuclear Threat Initiative, think perhaps the threat's personality traits need to be reframed: officials have long been focused on international extremists, but domestic threats—homegrown, often White terrorism—may be more of a problem these days.

No matter your nationality or political leanings, if you want to carry out an act of terrorism, atomic bombs are a good way to go: an explosion is relatively cheap on a per-death basis compared to other weapons. And, almost universally, people are more afraid of all things nuclear than they are of equally or more powerful conventional weapons. Then, too, there's the fact that a radiological or nuclear terrorist attack presents different political problems than would a nation-state's nuclear strike. Rogue actors are not directly tied to a government or even a location that the United States could retaliate against. Without the ability to launch a bomb back at someone who launched one at you, classical deterrence doesn't exactly hold up.

And getting the material might not even be as hard as stealing cesium from a hospital. A 2022 Government Accountability Office (GAO) report, for instance, told a disturbing tale. "Current assessments of the threat environment show an increasing interest in using radioactive material for making a dirty bomb," the report read. Investigators, curious how good the Nuclear Regulatory Commission would be at catching on to something nefarious of that sort, created shell companies and forged licenses with which to purchase highly restricted radioactive material.

They were able to easily order that material from two separate companies. Those companies took their orders, took their money, and prepared the material for pickup "by a representative of our shell company (actually, one of GAO's investigators)."

In the report's pictures, a box and a bucket with radioactive symbols smile at the reader.

The issue here is that purchasing and possessing these materials in these quantities would be perfectly legal—if the licenses and companies were legit. And yet an inadequate check of the latter could lead, in a different circumstance, to an abuse of the former: that legal, low-level material still could be repurposed into a weapon for domestic terrorism or used for other nefarious purposes. It wouldn't have been that difficult, because the hardest part of making a radiological device is getting the material. And it would have happened inside the country—not beyond its borders, where we often look for atomic threats.

The Nuclear Regulatory Commission bit back a bit, claiming in the report that the consequences of making a dirty bomb from the level of material the investigators had acquired "would be insufficient to require issuing immediately effective orders." And this even though such a bomb, the report notes, "could be expected to cause hundreds of deaths from evacuations and billions of dollars of socioeconomic effects."

If it did, NEST and NARAC would be there to help clean up and quantify the mess.

# CHAPTER FOURTEEN

Good luck following the evidence.

—*CSI: Vegas*

RESPONDING IN THE IMMEDIATE AFTERMATH OF A RADIOACTIVE DISAS-
ter or concern is only part of the work, though. Figuring out where the illicit material came from is a whole other discipline—and, unsurprisingly,
it has its whole own bureaucratic structure.

In June 2022, around seventy people—nuclear forensics scientists, reg-
ulators, and law enforcement officials from twenty-two countries and two
international organizations—converged on the Marriott in Pleasanton,
California. Amid the neutrally colored high-back chairs and hanging
disclike lights of the mid-tier hotel, this strange crowd—members of the
Nuclear Forensics International Technical Working Group—was be-
latedly celebrating the body's twenty-fifth anniversary.

First formed in the 1990s, the working group aims to advance nuclear
forensics, a field that's like *CSI* for scenarios involving nuclear and radiological material. During the meeting, they would be working through
such a scenario as if it were real. The exercise was slightly lower key than
Cobalt Magnet in that it didn't involve roving physically about a city. The
working group's task was more like playing Dungeons & Dragons than

LARPing—or what national security types more boringly call a "tabletop exercise." The goal was to improve their particular *CSI* skills: to identify a radioactive material and trace it back to its source.

Sometimes, after all, nuclear and radiological material gets lost, or stolen, or misrouted, or disposed of improperly. Often the material is part of a piece of equipment, and the person responsible for, say, a theft might not have even realized it was there. There was the time, for instance, in 2018 when a work truck containing a radiography camera was stolen from a gas station in West Virginia. On another occasion that same year, a radioactive seed being used as part of medical treatment was unaccounted for after it left the patient, perhaps having been flushed down the toilet with catheter contents. But any time stuff disappears, even if the disappearance itself isn't malicious, leaders rightly get concerned that it may not stay gone forever. Instead, it might enter stage left in an explosion, a trafficking ring, a terrorist cell, or the semisecret scientific establishment of a would-be nuclear country. If that happens, officials want the ability to trace the stuff, whether it be bare or within a bomb, back to its origin point. Categorizing and sourcing radioactive ingredients is the heart of nuclear forensics.

The field itself began several years after the Berlin Wall fell, according to a history cowritten by Lawrence Livermore National Laboratory's Sidney Niemeyer, a nuclear forensics pioneer and a speaker at the 2022 Marriott gathering. Immediately after the Cold War, an era with a curious mix of hope and tumult, a lecturer at Livermore posed a provocative question: "How quickly and with what certainty (if any) could we identify the origin of a nuclear weapon unclaimed by anyone?"

Niemeyer's boss, sitting in the audience, felt the query digging into him like a spur. And so, like any well-spurred scientist, he galloped toward an answer. First, he helped set up programs to understand the problem itself. By 1992, Niemeyer—who was then Nuclear Chemistry Division leader at Livermore—had gotten involved and tried to convince the other weapons labs to join.

The idea, though, got a sort of collective "meh" from Sandia and Los Alamos. And so Niemeyer took his concerns to a nonweapons

Department of Energy (DOE) lab, the Pacific Northwest National Laboratory. Together with its scientists, he published a paper titled "Attribution Assessment of Illicit Nuclear Materials," which suggested creating a new team that could bridge the divide between the scientists, who could do the strictly *CSI*-type work, and the (probably intelligence and political) officials, who would lead an overall investigation. Eventually, a DOE-sponsored conference led to the founding of the Nuclear Forensics International Technical Working Group.

That's a lot of meetings and lectures and papers to decide such a program was necessary, when in hindsight the need seems obvious: of course someone might want to abscond with radioactive material and use it for their own purposes. Maybe that purpose is simply reselling it for a sweeter price. Maybe it's an act of terrorism. Or furthering an internationally frowned-upon, illicit weapons program. The material could be cesium or cobalt—radioactive substances that won't go boom on their own but could become part of dirty bombs. More worryingly, the substance could be a significant quantity of uranium or perhaps plutonium— the stuff of traditional nuclear weapons.

Decades ago, though, such problems weren't at the top of officials' minds. Neither was the idea that if someone smuggled nuclear or radiological goods, scientists should be able to tell where the material had come from. Things changed around the early 1990s, though, when officials seized their first batch of illicit nuclear material in Switzerland and Italy.

Around that time, in 1995, the International Atomic Energy Agency started tracking when and how nuclear and radiological material has gone rogue—whether it was stolen or simply lost. The resulting database, still in existence today, pulls together incidents from individual countries' own reporting databases and systems on a voluntary basis. Within that data, it's estimated that from 1993 to 2022, material went missing more than 4,075 times, with 344 of those instances attributed to likely "trafficking or malicious use."

Three hundred is not three hundred thousand, but in this realm, one is a large number.

AT THE MARRIOTT, THE WORKING GROUP MEMBERS WILL PLAY OUT A trafficking scenario on their tabletops. In such cases, a helpful Power-Point presentation often lays out the basic narrative setup for them.

An exercise done in people's home countries and virtually through the Office of Nuclear Smuggling Detection and Deterrence, though not the one participants did at the Marriott, went like this: Authorities have received a tip from a confidential informant. A passenger at the Dallas International Airport might have a forbidden carry-on. Before he can board for his trip from Texas to Frankfurt, Germany, airport officials stop him. The presentation, at this point, shows an image of a man with his arms spread away from his body as a search takes place and authorities find his backpack. Rather than the traditional liquids, gels, and aerosols of most airport Ziplocs, the guy has a baggie containing an unknown black powder.

After taking measurements of the powder's properties, the team determines that the dark particles are uranium. In a picture, a radioactive symbol overlays the substance.

That black powder is the first mock sample working group teams will get to place on their tabletop once this presentation is over.

Investigators then, participants are told, travel to the man's Lone Star State house. It's a ramshackle, country affair, made of mixed materials and topped by a fading metal roof. Inside, they find a small metal cylinder—heavy for its size and, maybe unsurprisingly at this point, radioactive. It looks like a nuclear fuel pellet, like the kind used in power plants.

This is the participants' second mock sample. The third they'll receive comes not from the person or his pad but from the city of Frankfurt, where the man was headed. There, in an abandoned industrial facility, officials had seized uranium pellets five years prior. JPEGs in the PowerPoint depict gloved hands, holding out a tumbling batch of pellets—almost proffering them—near the graffitied warehouse.

The teams now get their hands on the three samples, field notes from investigators, and a report showing the presence of uranium-238 and radioactive potassium. With these, they have to answer some questions: Are any of these radioactive enough to be illegal? How should they be handled? How are they related to each other? Can they be traced back to a specific nuclear facility—perhaps the same nuclear facility? Were they made through the same process or in the same batch? Was it Colonel Mustard in the library with the candlestick?

LIVERMORE'S NAOMI MARKS SPECIALIZES IN CREATING TABLETOP EXER-cises like this, designing scenarios that are realistic, if rare, and giving attendees just enough data that the clever and dedicated can find the right answers. For each such game, Marks—part of the working group and Livermore's nuclear forensics team—makes a playbook, containing all the data and pictures of the scenario. She often takes the images herself, dressing her house up like a crime scene, accessorized with yellow tape. Her garage transforms into a fake lab. Sometimes she even puts on a Tyvek suit to playact an initial investigation and gets her kids to take portraits of her in staged scenes.

Marks is a good advertisement for getting into nuclear forensics in that she's an adventurous nerd who likes to have a good time and doesn't take herself or her serious job too seriously. She's also, for the record, a competitive speedskater, who says she admires the chutzpah of 2018 Olympic competitor Elizabeth Swaney.

In the 2018 games, Swaney—who hails from just up the road in Oakland—competed on the Hungarian halfpipe team. She had gotten in by doing a simplistic routine at lots of qualifying events, building up points by doing the easy moves perfectly. Once she made it to the Olympics, instead of performing flips and tricks, Swaney simply slid up and down the pipe, garnering last place. "I would just encourage positive vibes for everyone," she said on the *Today* show, addressing her critics.

It's a solid attitude to take into the nuclear complex, where it would be easy to have a bad time.

In Marks's lab, scientists characterize samples, trying to gain an understanding of their isotopes, trace elements, and shapes to figure out what they're made of and whether they're consistent with American materials. The department also maintains the Uranium Sourcing Database, which contains information about thousands of uranium samples from all over the world. If a rogue sample turns up, they can compare its personal characteristics to those in the database and see if they get a statistical match.

But after the forensics team gets such a sample and sends back an assessment to whatever mysterious federal entity wanted it, that's sometimes the end of Marks's road. "Depending on the customer, we don't always get to know whether we're right or wrong," she says.

THE WEAPONS LABS ALL TOIL ON SIMILAR TASKS AS PART OF THE government's broader National Technical Nuclear Forensics program. There's just one problem: there aren't enough people in the pipeline.

"Nobody grows up and thinks, 'I would like to be a nuclear forensic scientist,'" Marks says.

She sure didn't. But the field needs its next generation. And one way it attracts young researchers is through cosmochemistry: research on the compounds inside space samples. "People do like Apollo rocks," Marks says. They are, in drug parlance, a "gateway science."

Young people also come to the lab because of its equipment, which bests that of pretty much any university. "There is a greater density of mass spectrometers in this building than almost anywhere else," Marks says.

The need for new blood is true in forensics, and it's true across the nuclear complex. As of 2021, 40 percent of the National Nuclear Security Administration (NNSA) workforce was eligible to retire within the next five years. That presents a large problem for the agency's weapons program: a country that has not designed or tested nuclear bombs or manufactured plutonium pits in decades leaves its proficiency at those activities largely in the brains of graybeards—who will retire to Tahiti soon and leave the youngbloods with no institutional knowledge.

That's bad news for the United States, at least in the United States' opinion. Even in peacetime, of which the country has had essentially none, America likes to maintain a posture from which it could leap, at any moment, into a more active nuclear era: it never wants to be left without people who could design and make and test weapons—just in case.

Some believe that fields like nuclear forensics—despite its own lack of young workers—may be a way to draw more early-career scientists into NNSA's embrace and so fill gaps across the enterprise.

Forensics may not teach anyone how to blueprint a brand-new weapon, put together a shiny plutonium pit, or 3-D-print a new explosive. But it may attract people into the nuclear complex. Maybe these future nuclear workers are currently studying chemistry or physics. Maybe they just like problem solving. Maybe they have a sort of interdisciplinary itch that academia can't scratch. Regardless, in a lot of cases, they simply haven't heard of nuclear forensics or NNSA. Much of the agency's outreach involves introducing them to both.

To be attracted to the field, they just need to learn how their classroom-gained know-how could apply to this new subject, with its relevance to the real world. And maybe once they're in that real world, they'll come work at a national lab and drag down its average age.

In the view of Luther McDonald of the University of Utah, the worker gap exists not just because of the end of the Cold War and nuclear threats' subsequent recession from the public consciousness: it also exists because disasters like those at Chernobyl and Fukushima made "nuclear" a bad word in all contexts. "Both of those instances kind of caused a nuclear shock globally, as they should," he says. "People didn't want to do it." In other words, they didn't want to be a part of nuclear work. The sour taste extended to all applications of the atom-animated world, not just those with direct links to broken power plants. Partly to address the worker shortage, Congress passed the Nuclear Forensics and Attribution Act in 2010, mandating that the Department of Homeland Security set up education and recruitment programs to reel in students with relevant expertise. One such initiative, in place until 2017, was called the Nuclear Forensics Summer School. The six-week program rotated between colleges, including McDonald's.

The highlight was a grown-up field trip to a national lab partnered up with the host school. The labs are ready to give jobs to the early-career researchers who know what they're doing—or maybe just learned what they *could* be doing.

"They are really desperate," says McDonald.

MCDONALD UNDERSTANDS HOW THE SHIFT FROM BASIC PHYSICS AND chemistry—entailing the kind of fundamental research principles students are inculcated with in graduate school—to applied nuclear forensics happens and the reasons why scientists take the step across the threshold. After all, it happened to him. His path changed in his early studies of chemistry when he noticed something strange. "You talked about the entire periodic table except for the bottom row," he says. Those are the actinides—radioactive metals, always unstable, throwing around energy and particles in physics fits. Some, like uranium and thorium, are bound up naturally in the Earth, deposited in rock after stars birthed them billions of years ago. Some, though, are by their nature much newer: they are mostly made in the lab, by humans—who are more powerful, in this rare case, than the universe itself. That elemental group includes the mercurial element plutonium. McDonald, learning of these chemicals' complexity, became fascinated with their stretch of the periodic table.

Today his research focuses on uranium—how unique shape-based or microstructural attributes are introduced through chemical processes.

That study of actinide evolution is key to nuclear forensics. If rogue radioactive material shows up in, say, the Seattle airport, the work of researchers like McDonald can illuminate where the sample came from and how long it's been around. But McDonald's work is also useful from a basic-physics perspective: it demonstrates how uranium fundamentally—outside humans' woes about terrorism and dark markets and state war—works. In this way, nuclear forensics shrinks the already thin line between basic and applied science.

Whether McDonald is doing research for an agency like the Department of Homeland Security or one more like the National Science Foundation, the to-do list is functionally the same. "The only difference from my perspective," he says, "is I see more of what the end application might be" with the latter.

In a similar way, NNSA can capitalize on young scientists' interests in basic research to pursue national security projects. Ultimately, some of those scientists decide this particular applied focus—or maybe any applied focus—isn't for them, because of scientific or political or philosophical qualms or just because it doesn't light their intellectual fire. But also ultimately, a lot of young scientists end up finding purpose by doing work that has a direct application to and impact on the world, even if they didn't know that impact was part of a possible career path when they started.

The proposed pathway worked to introduce former University of California, Berkeley, student Eric Matthews to the field. As a freshman, Matthews started working with a professor named Bethany Goldblum on software called the Fission Induced Electromagnetic Response (FIER) code.

FIER could help nuclear forensics experts who study gamma radiation emitted by rogue nuclear material. It was also useful to chemists and physicists who just wanted to know how atoms behave in a more fundamental way.

Matthews began his education with that sense of curiosity. It was fun to learn about the way the smallest parts of the world work. "You also kind of pick up the attitude of your professors, and the attitude of a lot of pure physicists. It's a little bit arrogant towards applied science," he says. "It's like, 'Why even do anything that matters in reality?'"

Based on his phrasing, you can probably tell that Matthews doesn't agree with that attitude now, as a nuclear-data researcher. Today, he gets what seems to him the best of both worlds: basic science played out in experiments that he can then apply to reality.

Working in the field hasn't just changed his attitude toward science, though: it's also changed his views on weapons. When he was younger, he thought, "These things are bad. We should banish them, get rid of them completely," he says. Now, knowing more, he holds a grayer opinion, one shared by many of his nuclear elders and usually promoted by both the labs and their pipeline programs. "These things can't just be put away," Matthews says. "So how do you then manage policy that makes sure that they never get used in anger again?"

No one has the answer, really. And that's part of the reason nuclear forensics exists: it's a tool you might use in the aftermath of a nuclear incident. Here, then, as with stockpile stewardship and environmental management, this science exists to deal with previous creations and their messes and to yoke the next generation to that task.

# CHAPTER FIFTEEN

You never know: The next DJ Snake, the
next Skrillex, the next big DJs might wait
outside of the club. You gotta give back
and listen to the next generation and show
some love.

—DJ Snake

BETHANY GOLDBLUM, ERIC MATTHEWS'S OLD PROFESSOR, FOUND HER
way into the nuclear world in a similar way and has been bringing new
scientists in ever since.

Back when she was just a fresh college graduate, she planned to go to
law school, as upwardly mobile twentysomethings who want a second
degree are wont to do, even though she had studied chemistry and math
in college. But Goldblum changed her mind when she got an unexpected,
late-breaking note: she'd been accepted to the Nuclear Chemistry Summer School run by the Department of Energy (DOE). "That was really
where I decided I'm not going to actually go to law school," she says. "I'm
going to find something else to do."

That quest for "something else" started on the concrete next to
her grandparents' Louisiana pool. From that perch, Goldblum began

contacting her college professors. Maybe, she thought, they could give her a research project (and some money) once she finished the summer school so she could stay in the field. And she was in luck: one professor had just accepted a job at the Massachusetts Institute of Technology (MIT) and needed a student to make some samples to send to the International Space Station (as you do). That was cool, but it had just one caveat: "You have to get on a plane tomorrow," the professor told her.

"So I did," Goldblum says. "I just left."

Once she got to MIT, her professor's high-level position and her sideways association gave her a lot of latitude. To figure out what she might actually want to do with her life, she was able to ask other students to give her tours of their labs. "I just started traipsing around the institution and seeing what was there," Goldblum says.

The search for a suitable subject ended when she got to the nuclear engineering department. There, she stood near a strange device called a *tokamak*. The tokamak created plasma—a hot soup of charged particles, stripped from atoms, free and alone—and whirled it in a donut shape, which scientists sophisticatedly call a *torus*. Her tour guide told her he was using it to work on nuclear fusion. "I was like, 'Oh my god, that's so cool,'" she says. "And he was like, 'Really? Do you want to see my plasma?'"

The answer, of course, was "Duh." When he fired the machine up, it indeed produced a plasma; it was purple, and—this is key—it was his. To create and take ownership of such a wild thing represented to Goldblum a different kind of power and control than she'd ever had in the world.

That was when she knew she'd go nuclear.

Not long after, as a newly minted graduate student at the University of California, Berkeley, Goldblum began to doubt her path: she'd gotten a D on her first test. "Maybe I'm not cut out to be here," she recalls thinking.

But she was—she just needed to study more, or better, or both—which she did, upping her grade to an A by the end of the semester. Since those early days, Goldblum has become a fixture in both the nuclear science and nuclear policy worlds. She now works as a professor at the institution that gave her that D and as a researcher at the nearby Lawrence Berkeley National Lab.

Soon after Goldblum started graduate school, she knew she didn't just want to do science isolated from the outside world. It was the same moment when she realized that seemingly isolated science nevertheless impacts the real world, sometimes in the most literal sense of the word *impact*. The shift happened when she attended the Public Policy and Nuclear Threats Bootcamp, a workshop that she now, years later, is in charge of.

The bootcamp starts simple—answering questions like "What is a neutron?"—and climbs the intellectual ladder up to a rung like "What role should US nuclear weapons play in deterring conflicts in Europe?"

Within a week of starting the camp, Goldblum began having nightmares. In her REM-fueled hallucinations, nuclear weapons kept going off behind her rapidly shifting eyes: boom, boom, boom, boom, boom. "It was a realization of the policy implications of our technical research," she says. Prior to that, she'd thought she might go off and use her nuclear know-how to be a weapons designer. But when the dreams looped, her body paralyzed in sleep and helpless to stop them, she started to rethink that future. "The idea of going in there and building weapons without any mechanism for shaping the outcomes of the technical research," she says, "I just thought, 'I have to do something else.'"

But by the time she finished graduate school, she still didn't know what that something else would be. So she stayed on at Berkeley as a postdoc, as lost scientists in a competitive job market do, waiting for another moment that would change her life. Around that time, a department member applied for money from the National Nuclear Security Administration (NNSA) to start a program that aimed to snag nuclear science graduate students and show them what Goldblum had already learned at bootcamp: that their science isn't apolitical. But after that, importantly, it would teach them how to both understand and influence the way their seemingly basic, technical knowledge gets used in the real world. Today, Goldblum directs that program.

But back in 2011, it didn't exist, and NNSA's Office of Defense Nuclear Nonproliferation had just put out a call, asking for universities to come together and propose collaborative centers that would train the next generation of nuclear security experts. NNSA would provide $25 million to the winner, spread over five years.

Around the United States, would-be teams started forming, and several of them asked Berkeley—with its renowned nuclear engineering program—to join up. But Berkeley, being Berkeley, didn't go that way. "They decided that they were going to make their own," says Goldblum. It was the eleventh hour, uncomfortably close to the submission deadline, and many of the other nuclear engineering departments the school could have collaborated with had already been picked for other teams. "They decided that they would instead reach out to basic-science departments," says Goldblum. They could focus on educating young fundamental researchers about how their work might apply to nuclear security and nonproliferation. They called their group the Nuclear Science and Security Consortium (NSSC).

Scrambling to get the proposal together, they faced a last-minute speedbump: Berkeley wouldn't let the team submit to NNSA because its program had citizenship restrictions, and the school prided itself on its global openness. But if you know anything about nuclear, you know your passport is always relevant. The lawyers entered the chat and ultimately gave a green light at the very last minute.

Given the scrambling, the outcome was unlikely, but Berkeley's team won. It still exists more than a decade later and has trained hundreds of young scientists.


Today, recruiting young scientists who weren't even glimmers in their parents' gonads when the Cold War ended can be difficult for NNSA. Most students studying science only know about careers in academia—*be a professor like your professor!*—or in industry—*go make a lot of money creating products!* Working for a national lab, on topics that usually don't sit squarely within one scientific discipline, isn't often on students' radar. Half the battle, then, is just letting early-career researchers know what atomic options exist.

The DOE has created some programs to address this lack—funding fellowships for graduate students in relevant fields, for instance, and then bringing them to places like the Lawrence Livermore and Los Alamos national laboratories for internships and hobnobbing with lab scientists. Those fellowships don't usually have names like "Young Scientist Weapons Award" or "Early-Career Arms Designer Grant" but instead tend to deal with topics like nuclear safeguards (making sure radioactive material isn't where it shouldn't be, doing things it shouldn't be doing) or nonproliferation (making sure people aren't building too many bombs). One program, called the Nuclear Nonproliferation and International Safeguards Fellowship, targets doctoral students who might not know what *safeguard* refers to, besides a popular brand of soap, and exposes them to that world before they choose a career. NNSA also provides curriculum modules to university students.

Goldblum's NSSC provides a similar sort of opportunity, educating science students in the societal implications of their work and how to use their education and skills for good.

Goldblum, admittedly biased, sees NSSC's founding as a turning point in NNSA's aged-employee problem. "There was absolutely no plan for continuing to feed the pipeline," she says. Scientists were getting older, and the last of the generation who'd gone out to the Nevada desert to explode things were retiring, and also dying, their knowledge going to Tahiti or into the ether with them. But through NSSC, whose contract has been twice renewed, Berkeley and its collaborators have supported more than five hundred students, giving them money to live and work on research gigs at DOE facilities. "There continues to be a challenge in terms of getting the number of people that they need out to the labs," Goldblum says. "But I think that we've made a big dent in that."

To find its students, NSSC looks at the top applicants to nuclear-relevant science programs at its consorting schools. "Then there needs to be an overlap between their interest and the nuclear security field," she says—even if it's an overlap the students themselves haven't yet recognized. Maybe that's a student who is interested in neutrinos—ghost particles with almost no mass that travel at near light speed—but doesn't know that those strange guys can reveal information about the nature of radioactive material and help with nonproliferation, in addition to unveiling information about the state of the sun.

"From that crop of the best students, then we discuss with them, 'Hey, is this something that you're interested in exploring?'" Goldblum says. "And if so, then we bring them in."

It's a kind of soft-pitch approach, the nuclear version of getting someone to attend church by asking if they're interested in a cookie reception and kindness and, perhaps, a bit of tithing.

JAKE HECLA THOUGHT THIS SOUNDED LIKE A GOOD DEAL, ALTHOUGH HE was perhaps always destined to be a nuclear security researcher with a political mind, regardless of what incentive programs existed. Hecla got an early start, first discovering radioactivity in sixth grade. Back then, he was visiting what he calls a rock-and-junk shop in Roswell, New Mexico, while road-tripping with his mom. The store didn't just contain the usual Little Green Man stuff the city is most famous for (although it had plenty of that). It also housed a piece of uranium ore.

Hecla stared at the rock, scared, having seen what radioactivity had done to Bruce Banner (made him someone you wouldn't like when he was angry and also a *Big* Green Man). As Hecla and his mother drove on through the Southwest, he commandeered her BlackBerry to search online for words like "nuclear," trying to determine how likely he was to turn into the Hulk after having spent those minutes next to radioactive minerals. And as he continued to read, he grew—maybe paradoxically—less scared.

"From then it was pretty well set," he says.

His path toward nuclear engineering, his current field of study, began to crystallize around eighth grade, when he discovered an online forum called Fusor.net. It was a place where amateur researchers gathered to try to build a kind of fusion device called a Farnsworth Fusor. This fusor is, at least theoretically, so small and inexpensive that amateurs can build them in their basements and garages with components they purchase online. "Fusion is the process that powers our Sun and all the stars in the heavens. It is God's own way of powering the universe," says an explanation on the website. "Over the past sixty years, billions of dollars, pounds, and rubles have been spent trying to solve the riddle: how do you bottle a star? How do [you] contain a seething plasma as hot as the sun without either melting the container, or extinguishing the reaction?" Hecla was a teenager intent on figuring it out.

He wasn't alone in the quest.

On Fusor.net, Hecla found "a group of people who are retired or are not yet real adults, so they have plenty of free time." The old and the young had the most time to dedicate to the pursuit, and he, young, joined the strange crew. This period is still one of his favorite parts of his research career, which now involves professional labs at both federal facilities and universities. In actuality, he spent it in his grandparents' basement, scrounging around the internet for parts. "The moment that I knew I was able to do nuclear reactions with things that I built from eBay was deeply empowering," he says. It was the same sort of power hunger Goldblum initially felt, seeing a plasma spun up out of seemingly nothing. Although, of course, building a high-voltage device that emits X-rays presents a danger that operating a tokamak in a highly regulated university does not.

Hecla also spent his weekends with a group now called the Northwest Nuclear Lab, a sort of incubator for overachievers and their science fair projects, which Microsoft engineer Carl Greninger runs out of his own basement in Washington State.

The experience taught Hecla about machinery—and about how much he liked taking fundamental knowledge of small parts of the universe and turning it into a physical device. In college, he got his first job in the field. A professor needed a piece of lab equipment repaired and said if Hecla could fix the instrument, he'd get $15 an hour and coauthorship on resulting papers. Like a mechanic, Hecla got the machine up and running. In a chain reaction, this repair job led to more work, building X-ray systems for other older professionals, who lived more in their minds than in the details of the machines. Hecla, by contrast, liked to occupy both spaces at once.

Today, as a graduate student and an NSSC fellow, Hecla has been working on devices of his own—not just repairing those of others. In one project, he investigates the feasibility of using antineutrino detectors to monitor nuclear facilities' fuel cycles from afar—seeing, for instance, if reactors are off and on when people claim they are—and to find out whether nuclear facilities have only the agreed-upon ingredients in their nuclear reactors, not the sorts of materials that could be spun up into weapons. The antineutrinos—which are also ethereal particles that fly almost at the speed of light and hardly interact with other matter at all— emanating from labs and plants give away details of their inner workings. Antineutrino detectors, then, could reveal whether countries are building up plutonium in reactors. "I think it's incredibly ambitious," says Hecla. "It's kind of one of those wild, out-there ideas."

He's also spent plenty of time outside the lab, having traveled to Chernobyl, for instance, multiple times. On some of his trips, he's worked on a project called The Dogs of Chernobyl, which traps, neuters, and releases canines that paw around the exclusion zone and attempts to understand their radiation burden and its effects, on them and on the people that still work in that area.

Of course, he's the one who designed and built a device that determines how much cesium-137 resides within the canines to understand the persistence of the mess that was made.


HECLA HAS ALSO BEEN WORKING ON ANOTHER DETECTION DEVICE, alongside fellow NSSC member Kalie Knecht. Called Polaris-LAMP, it picks up gamma rays and uses them, along with other information, to make a 3-D reconstruction of a scene—and a map of what within said scene is radioactive, pinpointing sources of radiation and their levels. Unlike Hecla, Knecht wasn't always shooting toward a nuclear orbit. "I don't think that I wanted to do anything," she says. When she was a kid, she means, she didn't feel pulled toward any career in particular. But she always knew one thing: she wanted to move away from the West Virginia coal country where she was raised. And she has, about as far away as you can get in the continental states. No longer landlocked, she bought a sailboat while in school—for $500, a piano keyboard, and some uranium rocks Hecla had collected. She purchased it with the hope of living on it with her boyfriend, because the marina slip cost just $300 a month. Much cheaper than a Bay Area apartment. However, she noted later, "The saying 'there's nothing more expensive than a cheap boat' is true," and they're still working to make it livable.

She didn't really ever imagine she'd be here, bobbing in a bay, making pictures of irradiated parts of Earth. But her path began in high school physics class—that place filled largely with the imaginary incline planes and colliding billiard balls and force diagrams that bore most students. But among those banal topics, Knecht found something better: nuclear science. The power in atoms was so different from the fossil-fueled power plants of her youth. But as part of her education, she also learned about nuclear's second face: weapons. She hadn't even known they existed before that, even though the boys in her class apparently already held the opinion that "nukes are so cool."

Knecht went off to study nuclear engineering, intending to work at a power plant. It was, she thought, just what you did with that sort of a degree.

It paid well.

It was stable.

It pointed away from her past.

She wasn't very excited. But another event would bump her toward a different path. A faculty member took students on a field trip to visit the International Atomic Energy Agency (IAEA) in Vienna, tasked with furthering the "safe, secure and peaceful application of nuclear science and technology." Knecht learned about how experts there tracked and controlled nuclear material across the planet, using the same skills she'd learned in school.

That sounded more like it.

Today Knecht and Hecla work together as graduate students at Berkeley, on a project relevant to that goal, using Polaris-LAMP. With its gamma ray detector, location-pinpointing capabilities, visual camera, and onboard computer, it can paint a picture of a given place, in real time, and overlay a map of radiation in the area. Scientists and officials could use it to determine the source of a radioactive problem or to make sure, as the

IAEA likes to, that people in nuclear labs are only working with the material they claim, in the quantities they claim.

Trust but verify, with a device made by graduate students.

Those students, Knecht and Hecla, have schlepped Polaris-LAMP to the Paducah Gaseous Diffusion Plant in Kentucky, using it to survey containers of uranium products; to Fukushima; and, of course, to Chernobyl.

The images the device produces appeal to Knecht's artistic side. The regular 3-D map looks a lot like modern sonograms. Eerie, sketch colored, like someone made a detailed contour map using gray clay. The superimposed radiation ranges across the rainbow, in neon colors, emanating from hotspots and changing color the farther you go from the source. The combination ends up resembling a somewhat creepy airbrushed T-shirt. What Knecht likes most, though, is that the images tell a story—better than the staid sections of a journal article ever would. And sometimes the story's arc is less scary than that of popular narratives as the maps show emissions that wouldn't necessarily hurt you (much), even in places that remain nightmares in the public imagination. "It's a much easier way to communicate the facts and the actual risk when it comes to radiation, because there is so much general public fear about what radiation can do to people," Knecht says. "And I'm not saying, 'It's great, you should go bathe in a radium bath,' but I think there's a lot of misinformation out there."

KNECHT ISN'T SURE WHAT HER ULTIMATE CAREER PATH WILL BE. But many of the NSSC fellows, even if they'd never thought about nuclear security or policy before, end up staying in the field and in the national lab system—a commitment that tends to cement itself after they go work, as they all do, at one of the facilities during their graduate schooling. NSSC also gives students, outside their lab experiences, a way to marry technical research and policy. It's a web whose connections many tenured academics and career lab employees don't fully see, toiling away on their small problem, unaware of its big-picture implications or the avenues through which they might employ their device—a Polaris-LAMP, an antineutrino detector—out into the world beyond academic papers. "Many of them just haven't thought about their problem beyond 'Here's the math problem, solve the equation,'" Goldblum says. "And that's how it was for me approaching this too."

At the labs, employees do get schooled in some of that, but NSSC hopes to offer a broader philosophical perspective. "A lot of even the lab education is taught from a kind of 'drink the Kool-Aid of deterrence' view," Goldblum says. "And I'm not saying that there's anything wrong with that perspective, necessarily, but I think you need to hear other perspectives."

That's why, as part of the Public Policy and Nuclear Threats Bootcamp that students attend, Goldblum brings in speakers with various ideologies. In some ways, many of their ideas are just different flavors of the deterrence Kool-Aid. Some think modernization should involve building new weapons with new capabilities, while others think modernization should just fix the degradation of the same old bombs. Some think the United States needs an overwhelming advantage (more blue-ribbon bombs), while others like to contemplate how few the country could have and how feeble while still warding off attack. Still others think there should be no nuclear weapons at all.

You can hear some of that discussion at the labs, but in the end, the employees are there to play out the literal party line, passed down by Congress—not to choose themselves, say, where plutonium pits should go and who, if anyone, should make them. The NSSC students' exposure to various ideas before fully jumping aboard with "the mission" is meant to give them the tools they need to craft their own views.

"If you just hear what they're saying and do blindly what they say," says Goldblum, "that's not good."

Goldblum doesn't like to share opinions on specific nuclear action items. But she feels like modernization and pit production are straightforward enough that there's not much room for philosophizing. If nuclear weapons are going to continue to exist, which she believes they will, they can't stay as is—at least not if they are to remain safe, secure, and reliable, a phrase she uses, just like everyone else does. "I can feel very comfortable in saying that there's no doubt that this is a required thing to do," she says. People can debate, as they do with the NSSC fellows, whether the United States should develop new capabilities and where and when to make how many pits.

"But in terms of whether or not we need to modernize," Goldblum says, "there's no question in my mind."

# Part Four

# BOOM TOWNS

THE TOWNS IN WHICH THE NATIONAL NUCLEAR SECURITY ADMINIS-tration labs situate themselves (and the ones they have taken over) often have a complicated relationship with the nuclear complex. The ways lab employees think of themselves don't always mesh with the view from the outside.

Neighbors, the general public, and even nuclear policy analysts don't see the work of the lab the same way those inside do. And that's true even when the disparate groups have the same aim: a safer world.

# CHAPTER SIXTEEN

There are three ways to encourage initiative. One is to cut off people's heads
as they do in Russia. Another is to subject people to public criticism, which is
impossible in such secret work as this. A
third way is to set up competition. This is
Livermore's most valuable function: simply
to be a competitor.

—Edward Teller

JUST SOUTHEAST OF BERKELEY, WHERE BETHANY GOLDBLUM IS HELP-
ing raise up the next generation of policy-minded nuclear scientists,
is a town where that policy gets put into practice: Livermore, home to
Lawrence Livermore National Laboratory. When activist Marylia Kelley moved there decades ago, she hoped it would be a bucolic place to
raise a kid. Sure, she knew there was a government lab in town. She
also knew lots of residents seemed to work there. But she didn't know,
at least not at first, that the lab existed largely to work on nuclear weapons. Nobody talked about that work, and at the local newspaper—where
she worked part-time—no one wrote about it either. When the lab and

outside commentators did discuss Livermore, they tended to stick to its more PR-palatable pursuits, which today range from designing satellite sensors to furthering fusion energy. "Eighty percent of the press releases are about 20 percent of the budget," Kelley says.

The other 80 percent of the budget, of course, goes to bomb stuff.

Livermore, in character, is different from Los Alamos. The town doesn't necessarily derive its identity from the lab—or atoms. There are no atomic symbols on its trash cans. And the biggest hint in May 2022, besides security gates and a few signs, about what goes on takes the form of a street banner advertising the Fusion Soccer Club.

Livermore has the feel of a California suburb, which it pretty much is, with its low-slung strip malls and its quaint historic downtown—hosting restaurants, coffee shops, salons, and boutiques. Neighborhoods form the edges.

That philosophical distance makes sense since Livermore as a city wasn't founded specifically to create bombs. Since long before that purpose descended upon the town, Livermore's rolling yellow hills, which turn green in winter, have been home to prosperous ranchland and one of California's oldest wine regions. Livermore, the lab, was founded after World War II, at the former site of a naval base, as a sort of rival to Los Alamos—a competition that would spur better weapons work from both facilities. Early on, its primary goal was to help make a useful hydrogen bomb—a goal that J. Robert Oppenheimer opposed.

Some lab employees say Livermore has that small-town, everyone-knows-everyone feel, where your coworkers might eavesdrop on your loose brewery conversation. But the strength and truth of that feeling are relative: the town still has little of the insularity of Los Alamos. It's part of the East Bay Area, not a secret city on a hill. Perhaps because of this and its residents' integration with the outside world, its scientists tend to speak in fewer nuclear aphorisms. You won't hear as much "always, never" talk, and while people speak often about the concepts and rationale behind a "safe, secure, and reliable" deterrent, the phrasing gets more individual variation rather than having the feel of an incantation.

The lab seems to conceive of itself as a more laid-back but sophisticated version of Los Alamos—less militarily hierarchical, more worldly.

And that is perhaps true. But drive a bit outside town, on the two-lane highway where people pass on the double solid line, and you'll find a rougher part of the campus: Site 300. At this seven-thousand-acre spot, Livermore does its large-scale, high-explosive testing—minus the nuclear material.

The site isn't secret—you can find it on Google Maps—but it's not exactly well advertised to the general public either. The only real clue that something interesting might lurk up the road, as you leave Livermore proper and head into the hills, is a suspicious number of white trucks, all the same brand and size.

At an entrance to Site 300, there's a clear parking area, a fence with Department of Energy (DOE) warning signs, and a perimeter road. But no signs along the main road direct people to the site or suggest its existence. It's nestled in lumpy tinderbox hills, which don't look like a great place to set off a bunch of explosives. But out of view of the roadway, behind the fence, black burns scar the lab landscape.

Once Kelley had done more research about the work happening in her backyard, she couldn't let the knowledge go—or keep it to herself. People deserved, she thought, to have the same information she had and whatever further information she could get. "My feeling was they can make up their mind," she says. "But you have to know before you can make up your mind."

The lab, in Kelley's view, robs people of that knowledge by keeping its activities secret—or at least quieter than it is legally obliged to. "Salute us and don't ask any questions," she says of its employees' attitude.

"But," she adds, "I have questions."

Back in the 1980s, her eyes opened, Kelley visited a "peace camp" outside the lab. After the activist occupation was over, she sat around a kitchen table with like-minded neighbors and imagined the environmental impacts of a lab like Livermore and the Sandia outpost that sits next to it. "We'd never heard of any," she says. But Kelley, whose background is in journalism, knew how to wield the Freedom of Information Act and the California Public Records Act. Both legal maneuvers gave her the details she needed. "Once we got ahold of that information, things were much worse than we had ever imagined," Kelley says.

The released files reportedly contained, for instance, an image of two hundred unlabeled, leaking drums, sitting on asphalt, and a structure called the "Unknown Chemical Trailer." Researchers would leave in this trailer unlabeled, thus unidentified, materials that they wanted out of the building, according to Kelley. The environmental and human health impacts of those substances concerned her, as did the apparent lack of oversight of or responsibility for potentially hazardous materials—not to mention the uncertainty of material without a known identity.

Kelley and other rightly concerned residents formed a watchdog group, calling it Tri-Valley Communities Against a Radioactive Environment (Tri-Valley CAREs, or TVC). The term *tri-valley* is supposed to convey that the organization's membership comes from the community surrounding the lab, not "that bizarre place," says Kelley, referring to how the Livermore leadership might see nearby places like Berkeley and San Francisco.

TVC investigates the health and environmental effects of the lab's activities. "These things were not really known in 1983," Kelley says. "Nor did we realize that we were actually part of a citizen movement happening all around at about that time."

Antinuclear activism was in fact hitting a peak in the 1980s, and TVC's structure—a nonprofit watchdog group comprising those local to a national lab—is now replicated at other National Nuclear Security Administration (NNSA) facilities.

TODAY, TVC HAS SIX THOUSAND MEMBERS, INCLUDING EMPLOYEES OF Livermore and a California outpost of Sandia. "When you get down to the individual-scientist level, the individual engineer, the individual support person—there's a great multiplicity of feelings and ideas," Kelley says.

Some aren't pro-disarmament but appreciate TVC's existence anyway, from a waste, fraud, and abuse perspective. The organization's more conservative members simply believe the lab shouldn't be pursuing sky-high projects with little chance of success.

Kelley sees pit production as one project that shouldn't be going forward as is. Even though Livermore is not itself creating any pits, NNSA will send plutonium to the California lab, which will require transporting hazardous material across the country, to the very populated area of Livermore. And before doing that, the DOE isn't required to do an environmental impact statement evaluating how the final pit plans might affect Earth and its communities. In other words, the lab, in her view, hasn't done its homework to lay out the consequences of this change.

In light of that, TVC filed a lawsuit, suing the DOE and NNSA in an attempt to force the agencies to do that environmental work. Joining the California group in legal action are its cousins from across the country: Nuclear Watch New Mexico and Savannah River Site Watch.

ALICIA WILLIAMS, WHO WORKS AT LIVERMORE, DOESN'T SEE NUCLEAR modernization the same way Kelley does. But, of course, she wouldn't: she leads the W87-1 engineering program. The W87-1 is a modernized weapon to be added to the arsenal, meant to replace the aging W78 warhead, the first units of which were produced in 1979. It's a modification of an existing design, based on the W87-0—itself an old weapon, initially manufactured in 1986—that will now have more safety features and updated explosives.

This program is not quite as controversial as the development of a bomb called W93, which is a brand-new weapon—something the United States hasn't made and fielded since the Cold War. Creating a novel weapon could be seen, internationally, as a provocative move, critics say. Domestically, it's an affirmation that reliance on nukes will continue.

When Williams was growing up, that reliance didn't seem like it would last. She recalls hearing that nuclear weapons existed but that we wouldn't require them. And yet here she is today, in an office where she helps design an improved weapon.

NNSA, for its part, maintains the W87-1 has no new capabilities and is coming into existence only because of a desire to make nukes more safe, secure, and reliable.

Williams, who has one sticker on her wall that reads, "I ♥ requirements," and another that reads, "I'd rather be writing requirements," clearly falls into the NNSA camp. Work on the W87-1 is, in her mind, key to deterrence—to stopping, not starting, nuclear war. "The countries that I would not like to call our adversaries but seem to be anyway are, in some cases, doing technological things that are frankly breathtaking," she says (she does not name names or specify things). "They're investing a lot."

To make the tit for tat of deterrence reliable, in her view, we also have to try some new things. "It is an interesting juxtaposition," she says, "working on an incredibly destructive weapon so we never use it in anger."

After Williams moved to California but before she started working at the lab, she had barely heard of it. "That's a Department of Energy lab," she remembers thinking. "They must do science."

When she learned, as Kelley had, about what the lab actually worked on, she felt differently. Her work at a company that made endoscopy products seemed like it didn't have much effect on the world. "It just seemed so small," she says. The lab's work—upholding deterrence by maintaining the deterrent—loomed large.

"It means something," she says, sincerely.

But it's hard to alter and modernize a weapon when the United States hasn't done much similarly creative nuclear work in decades and so few people have the right experience. "There's not a deep bench of people," Williams says. The team she's been helping to build—which began with just her and a few others—had three thousand staff in March 2023, half of them five years or fewer out of school. With this new generation, Williams hopes to make a safer, more secure weapon than the one it's replacing.

Asked what sorts of things make the W87-1 safer, securer, and more reliable than its predecessors, Williams's colleague, Dan Haylett, punts. "That's probably where I can't go," he says.

The secrecy makes it hard to do anything but take his word for it—or your own word against it.

Not all of the steps of modernization are secret, though. As usual, the nonnuclear aspects are more open. And that's why the public can know that the W87-1 uses a different kind of explosive than its predecessor.

Historically, weapons had conventional explosives at their cores: they're touchy and eager to go off. Since those earlier weapons were made, though, scientists have developed "insensitive" high explosives. These devices can be exposed to heat or shot at by guns: no matter what, they won't ignite unless specifically triggered to detonate. That's ideal for nuclear weapons, if you indeed never want them to go off when they're not supposed to.

Livermore researches these substances in its High Explosives Application Facility. The building is unnavigable, a maze of rooms with strangely colored tiles that, it turns out, encode how many grams of explosives you are allowed to take into which areas. The hallways have names like roads—Chemistry Row, Trinity Parkway—so people don't lose their way.

Deana Kahnke, the facility manager, is proud of the experiments they do in this part of Livermore. The lab studies, in part, the differences between the two types of explosives, based on how they behave inside test chambers. These firing tanks—the smallest meant to handle one kilogram of explosive material, the largest, ten kilograms—are spherical metal containers that open in the middle. They look like diving bells, live in rooms with blast doors, and are reached by winding hallways whose twists are meant to divert explosions' power. Near one, a handwritten note stuck to a computer says, "Not a shelf."

The ten-kilogram tank is room sized and puts on a good show. Actually, more than a thousand shows: an American flag bow hangs from the tank's control room door to commemorate the fact that it has done 1,776 shots, or detonations.

Once a tank is closed and the scientists are safely behind secure doors, they can simulate harsh conditions and see how explosives react or just try to explode them and see what happens.

Kahnke shows off some results from a recent test, in which scientists confined explosives in a metal cylinder and heated them. The sensitive explosive, predictably, made shrapnel of its cylinder. The insensitive material, meanwhile, simply began to break down. As it did so, it released gases that split its container, but its core material didn't go off. Success.

Safe. Secure. Reliable.

# CHAPTER SEVENTEEN

The United States won the Cold War, not on a battlefield in some far off place, but in the Savannah River valley of South Carolina, the isolated deserts of New Mexico and along the Columbia River in Washington state. Those are among the dozen places scattered across the continent where America created and built its nuclear arsenal. Today, the nation continues to cope with the legacy of that creation.

—Editor's note, *Post and Courier*

ACROSS THE COUNTRY IN SOUTH CAROLINA, ACTIVIST TOM CLEMENTS had a similar realization to Marylia Kelley, although he didn't take organizing action till years later. During graduate school at the University of Georgia, he was surprised to learn that spent nuclear fuel might find itself rolling on trains through his college town of Athens, headed for a planned reprocessing plant in South Carolina—a private facility next to a place then called the Savannah River Plant. Only after he learned this information did he grasp something larger: the nuclear world has outposts spread nationwide. "There's this complex around the country scattered around like military bases," he says. It's a fact most Americans are scarcely aware of.

In what became Clements's backyard, the National Nuclear Security Administration (NNSA) has already considered enacting similar plutonium plans to those it's successfully starting up now. In 2005, a program to create a Modern Pit Facility, for which the Savannah River Site (SRS) had been under consideration, was canceled. After working with Greenpeace, Greenpeace International, the Nuclear Control Institute, and Friends of the Earth on issues related to the Department of Energy (DOE), Clements moved to Columbia, South Carolina, and founded the watchdog organization Savannah River Site Watch (SRS Watch) in 2014 to formalize his ongoing opposition to nearby nuclear work.

"The proposal to build a pit plant at SRS has been sleeping," he says. But it's extremely awake now. And SRS Watch is trying to stop it.

"I call it the SRS Plutonium Bomb Plant," says Clements, who runs the organization as a sort of one-man show. He would prefer, first, that pit production not happen at all and, second, that it not happen in South Carolina. In his view, it's neither good nor necessary, and the national security motivations are faulty.

"The lifeblood of the DOE complex is new capital-intensive projects," says Clements, agreeing with Kelley. Big programs that will have big budgets for a long time. "They finish one, they come up with another."

Pit production, in his view, is another.


THE SAVANNAH RIVER SITE, WHERE NNSA WILL BE MAKING FIFTY PITS per year in the 2030s, is a different beast from the other NNSA spots. It's accurately called a "site," not a "lab," for instance, because it is a place where things happen, not necessarily where things are investigated. It has the feel of a factory—albeit one that fears everyone might steal its materials, proprietary information, and secret recipes.

When you drive toward the site's main gate, both sides of the road are lined with trees of suspiciously similar heights, nearly forming an arched canopy above the cars. It looks like you're traveling to a trailhead. And that perception is almost accurate: NNSA operations use only 10 percent of the 310 square miles of land that SRS owns. The rest is managed by the US Forest Service and "harvested" every year for lumber. The trees help with erosion, which is the first thing Bob Bonnett, the site employee leading a private tour, mentions. But they also help with secrecy, their leaves shielding the site from view.

The trees and natural emptiness of the area, Bonnett says, also mean that if there were an attempt to infiltrate or disrupt the facility, the bad guys would have a hard time hitting everything all at once or traveling between buildings without getting caught. But he says less about the fact that the spread also somewhat protects the site's reactors and other processing facilities from each other—from each other's accidents.

Through the trees, you can't see the forest of infrastructure until you approach the badge office, where you'll be greeted by private security contractors from a company called Centerra. Its guards run the trails on-site to keep in shape and ensure they can chase any trespasser down. They are dressed like soldiers, with paramilitary weapons and demeanors, and demand that you leave your backpack in the car, like paranoid convenience store clerks. Twelve canine teams aid in keeping the site and its dangerous materials secure.

SRS's construction was initiated in late 1950, and the site got the catchy nickname "the bomb plant" soon after President Harry Truman announced his decision to construct it. That moniker was partly thanks to the *Augusta Chronicle*'s front-page articles titled "$260,000,000 Plant Means Big Boom for Augusta" and "H-bomb Material Facilities Will Be Constructed in SC." The president intended the land to be a place where workers could manufacture substances like plutonium and tritium, a radioactive isotope of hydrogen that helps amp up a fusion weapon's chain reaction. The people who work at SRS don't love its nickname because the site has never actually housed a fully assembled weapon—just the materials for many weapons—perhaps a distinction without a difference to those on the outside.

The bomb plant's postal address is in Aiken, which is a sprawl of sub-urbia about half an hour from Augusta, a midsize city. But the site is actually closer to the small, rural town of Jackson, South Carolina. It doesn't matter much either way, though: SRS is a city unto itself, larger than the interior of the beltway around Washington, DC. Aiken was a good spot for the facility because South Carolina was beyond the USSR's bombing range; it was near the Savannah River, which could provide a supply from which to make heavy water for the nuclear reactors; and its relatively mild climate allowed year-round work. There was just one hitch: around six thousand people lived on this ideal land at the time the plant was to be constructed. No problem: the government removed the people, eminent-domaining away their farms and their entire towns. Their agri-cultural spreads were replaced with the aforementioned trees and also a whole lot of concrete infrastructure: square, off-white buildings spread across SRS's metro area.

Today, three of the Cold War reactors on-site have been decommis-sioned and are now what Bonnett calls "huge blocks of concrete." They look like bunker cities in an apocalypse movie. "We were built for pur-pose and not for beauty," notes Bonnett. The other two reactors got retro-fits so they could store materials.

Up until SRS actually starts making plutonium pits, the facility's focus won't be on production. Instead of making new tritium, SRS recycles it from existing weapons and extracts it from old nuclear reactor rods. Much of its work is actually in remediation. A lot of effort at the site goes toward cleaning up messes from yesterday—some homegrown, others grown elsewhere. SRS processes and stores excess plutonium and works to chem-ically reduce and dispose of radioactive waste. Yet, because the United States has no permanent nuclear waste storage facility, what was intended as a temporary solution has become semipermanent.

Today, SRS holds tons of weapons-grade plutonium on-site—plutonium the DOE was supposed to have removed by 2016. The US government paid the state of South Carolina a $600 million settlement and is legally obligated to remove the plutonium by 2037.

The local *Post and Courier* has called SRS one of the most contami-nated places on Earth. And it has reported on workers who were exposed to unsafe conditions, developed illnesses, and even passed away. "Thousands of sick, dying and dead workers from the Savannah bomb plant and the nation's other nuclear weapons facilities manned the front lines in America's fight to win the Cold War," the newspaper said in a 2017 article, "and they are among its only casualties."

SRS has not, it's true, had the squeakiest safety record. It's played host to dangerous reactor accidents. Drums of contaminated dirt from elsewhere have been shipped to SRS and buried on-site (the radioactive landfill was later capped and closed). The site has also buried its own waste in trenches. Contamination has sometimes leaked into the groundwater. On a side road, a gate notes "Asbestos pit closed." Given all of this, it's perhaps no surprise that SRS is a Superfund site, which the Environmental Protection Agency says has risks that include "radioactive isotopes in the environment (sediment, fish, deer, soil, and groundwater) such as Cesium-137, uranium and tritium."

Perhaps unsurprisingly, radioactive alligators reportedly live in SRS's ponds, according to the *Post and Courier*, and radioactive cockroaches click along the ground. Signs along the campus roads warn periodically of underground radioactive material. SRS has been cleaning up its own mess, and those of other sites, for years, but there's still toxic stuff—and still concern. Multiple times a year, for instance, the site hosts a deer hunt to cut down on overpopulation. Around one thousand people come—but their deer carcasses have to be scanned for radioactivity before the bounty goes home.

This general contamination is why researchers take thirty-five thousand environmental samples a year, partly through the Savannah River Ecology Laboratory (initially known as the Laboratory of Radiation Ecology) to understand how radioactive chemicals have altered the planet.

SRS's CULTURE IS QUITE DIFFERENT FROM THAT OF THE RESEARCH LABS. On-site, public service signs give away the gist: it's heads-down action and high-security mind-sets here. There are no inspirational innovation-lab billboards, though one does say, "Starve a spy, feed a shredder." A church-style letterboard proclaims, "By failing to prepare, you're preparing to fail." A hallway notice announces, "It is hunting season, and Centerra is hunting firearms in your vehicle."

Talk touches little on deterrence and more on just getting the job done. As Bryan Cox, a site spokesperson, notes, the whole ethos orbits around the site's historical goal of sending shipments of tritium to the Department of Defense on time. It's a production culture, passed down through generations of southern SRS workers. While the other labs tend to bring in more outsiders, snapping them up from postgraduate programs across the country, in South Carolina local kids grow up aiming to work for the NNSA site next door. It's a good, reliable gig—with wages, Cox claims, twice those typically available outside the fence. "If you live here, getting a job here is the ultimate," he says. And there will be more jobs as pit-production preparation begins in earnest.

In certain circles, there's resistance to making the pits at all. But some (particularly those in New Mexico who don't want production to happen in their state) think that if you're going to do it, SRS is the place because of the factory-minded "deliver the shipment" focus. That on-spec, on-time mentality, though, may only be aspirational, or better applied to established missions with well-formed processes and infrastructure. Because it hasn't necessarily held for new, large-scale projects of late. Just look at the Mixed Oxide Fuel Fabrication Facility (MOX). That SRS program was supposed to turn weapons-grade plutonium into more benign reactor fuel. But after it was projected to be billions over budget and years behind schedule, the program was canceled. The giant concrete shell of a building constructed for MOX is now the future home of pit production. And so a site once meant to dispense with the plutonium will now be shaping it. Just as some people thought nuclear weapons might be on their way out after the Cold War, only to watch them come roaring back, this facility will be building up the nuclear complex instead of deconstructing it.

Perhaps unsurprisingly, the SRS part of the pit project is currently set to be around five years behind schedule, and in October 2022 officials revealed that approvals for the facility's construction would be delayed by around six months longer than they'd expected.

WHILE BONNETT, THE FACILITY GUIDE, MENTIONED THAT THE CAST-aside reactors resembled something out of *The Walking Dead*, he did not say the same about the new pit-production facility. Nevertheless, it too has distinct zombie-apocalypse vibes. It's hard to get a good view of it from online images, because they all seem to be taken from above. In those shots, the former MOX plant has a snub-nosed-gun shape. Seen in person, the sides of the three-story gray building have striped stains.

Getting pit production going here is going to test the mettle of the site and its suppliers. Making the building's exoskeleton into a full body is the largest part of the project, and the construction process will involve around twenty-five hundred employees. The gloveboxes that technicians use to handle that plutonium might not be available at the necessary scale or on the necessary timeline. Workers can't handle the substance without that equipment, so the program can't get into full swing without it.

Once building is done, though, someone has to work there—and that part of the project will require eighteen hundred people to support it. To help with that, the Savannah River Site and Los Alamos National Laboratory pulled together the so-called Knowledge Transfer Program, through which South Carolinians go out to New Mexico to learn about pit production, then come back to South Carolina to work on its production program and train others.

Near the future home of pit production, there's another building in which that training will happen. It's a mockup of what workers will encounter in *The Walking Dead* building, but with "cold" surrogate material instead of plutonium. Technicians will get qualified to do all the tasks in the cold setting before they're allowed to "go hot"—a process that officials estimate will take four or five years for each worker, not dissimilar from the time it takes to finish graduate school. All to make the pits that critics insist aren't necessary, that proponents say are critical to the future of the country, and that people on the outside can't know too much about.


OVER IN NEW MEXICO, JAY COGHLAN IS ONE OF THE CRITICS. HE CALLS the pit program partly "makework." Coghlan, head of a watchdog group called Nuclear Watch New Mexico (NukeWatch), has strong feelings about the role that the nuclear complex plays in his state. "Basically, New Mexico is number one in nuclear weapons and radioactive waste," he says, "and just up from the bottom in citizen well-being and child well-being." It's the lowest-ranked state in education and one of the nation's poorest. Meanwhile, in 2023, the DOE spent around $9.4 billion there. That's more than the operating budget of the state itself. "New Mexico is an extremely beautiful state, and I wouldn't live anywhere else," Coghlan clarifies. "At the same time, it's deeply troubled."

But Coghlan and NukeWatch—along with other watchdog groups like the Southwest Research and Information Center and the Los Alamos Study Group—have occasionally been successful in trying to counter that trouble. In 1998, for instance, they settled a lawsuit—filed along with dozens of other groups—against the DOE for "failure to provide adequate environmental review of its plans to clean up the mess created by more than 50 years of nuclear weapons research and production," according to the Western States Legal Foundation. The settlement required the DOE to create a public database of information about hazardous and radioactive weapons waste and provided $6.25 million "to assist community groups and tribes across the country in assessing the complex technical issues surrounding the ecological and health effects of nuclear weapons activities." That money funded organizations like Coghlan's for a while, but, in general, he doesn't feel he's had the impact he'd like.

"I draw blood," he says, "but I can't kill the beast."

# CHAPTER EIGHTEEN

I'm interested in the man and what [invent-
ing the atomic bomb] does to the individual.
The mechanics of it, that's not really for
me—I don't have the intellectual capability to understand them, but these contradictory characters are fascinating. . . .
People identify with that, because we all
walk around with these contradictory ideas
coexisting in our heads.

—Cillian Murphy, on playing
Oppenheimer in *Oppenheimer*

THE BEAST IS DOING WELL IN MARCH 2022. IN LOS ALAMOS, HOLLY-
wood types are milling about just up the road from Jay Coghlan, film-
ing the movie *Oppenheimer*—a biopic about J. Robert Oppenheimer and
the Manhattan Project. NBC Universal trucks sit idle by physicist Hans
Bethe's old house, which is now a museum. Inside the house, there's a
real Nobel Prize, and a sign on a bathroom closet says, "Please do not
open—unless you know how to close it." It marks the door of a potential
Pandora's box, apparently, which is appropriate. Potted ginkgoes grow

there, sprouted from seeds that Hiroshima's trees dropped after surviving the bombing of that city.

In Bethe's backyard, though, the film crew is busy building a forest of a different sort, lining up trucked-in trees with a much less complicated history, their needles obscuring the view. The movie is making the whole town into a set. Inside Fuller Lodge, which was a community center of sorts during the Manhattan Project, the informative signs about what *actually* went on here have been stuffed into side rooms to keep the set clear. The Manhattan-era truck that previously sat outside Oppenheimer's old place has been moved aside to make room for shiny new ones, presumably belonging to the film crew.

But despite this erasure, the town itself always remembers its past and revels in its glory days, whether that remembrance is obscured by a film crew or not: fraternal statues of Oppenheimer and General Leslie Groves, director of the Manhattan Project, stand watch, facing each other as if flirting, in the center of town. At the Bradbury Science Museum, the introductory movie focuses on the Manhattan Project, relegating the present to the exhibit hall.

The lab's influence, past and present, doesn't stop in town: you can drive for miles and still be surrounded by its property. Historical buildings and antitrespassing fences peek out of spaces that, in another region, would just seem like hiking areas.

Near Bandelier National Monument, an antenna looms on Department of Energy property. But long before that dish was here, ancient people dug cliff dwellings into cliff faces and community gathering spaces out of the ground, both now memorialized with interpretive signs and a guiding brochure. These earlier inhabitants left their own mark on the land, one less messy than those of today.

THE LAB IS CURRENTLY EXPANDING WHAT IT WILL LEAVE BEHIND FOR future archaeologists, something that Los Alamos National Laboratory (LANL) director Thom Mason addresses at a town meeting in June 2022.

When the virtual gathering opens, Mason sits in a room empty except for LANL communications team member Joe Gonzales, his foil.

Wearing tan pants, a light blue shirt, and a jacket, Mason is positioned like a talk show guest. He is half facing the camera and half facing Gonzales, ready to give a PowerPoint presentation on the latest at LANL and take questions from community members.

The lab, Mason began, is hiring twenty-five hundred new staff, its budget growing by $1.5 billion. "It's not about growing the size of our deterrent, or adding new capabilities," he says. "We just have to make sure that the deterrent we have is safe, secure, and reliable."

After the Cold War, he continues, it was possible things—on the planet and up on the mesa—would be different. "There was some hope that we would be headed to a world where nuclear deterrence would be less important," he says. "And that's not the world we find ourselves in in 2022."

He is speaking, in particular, about Russia's invasion of Ukraine. "What's the meaning of deterrence in this context?" he asks.

"To attempt to limit that conflict," he continues, answering himself.

What he means, he goes on to explain, is that nuclear deterrence is keeping Russia out of North Atlantic Treaty Organization countries. If you asked Ukrainians, though, they might say that Russian nuclear weapons are allowing the conflict to continue within their borders.

The United States is—and Thom Mason, head of a nuclear weapons lab is—affirming the continued and perhaps increased importance of a nuclear arsenal. The stance supports the lab's continued modernization and production work. "If we simply wait until those pits age out, we will need another Rocky Flats," he warns, calling up the biggest specter in the history of pit production, "which is a much larger proposition, much more expensive, and really not what we need in a world where we don't have 30,000 weapons."

Even for community members for whom that's not *philosophically* complicated, it can be *practically* complicated. Los Alamos has a limited housing supply, and what does exist is expensive. The town wasn't built to handle thousands more workers. The roads weren't built to handle thousands of extra commuters. People are concerned about a lack of child care.

To help with some of this, LANL recently leased space in Santa Fe for hybrid workers. That way, they don't have to commute, and they can spend their lunch money in the city's downtown—a place that's actually part of Los Alamos's roots: Manhattan Project workers had to first show up at a nondescript office at 109 East Palace in downtown Santa Fe to be vetted and then directed to the lab.

But expansion into Santa Fe is exactly what some are worried about. In early 2020, protests organized by the Los Alamos Study Group came together beneath the city's statue of St. Francis of Assisi—patron saint, sometimes, of animals and the environment. The three dozen naysayers opposed the lab's bid to develop part of the city's midtown. Protesters' signs shouted phrases like "No war work in Santa Fe," "City of faith or city of nuclear weapons?" and "LANL is not an education institution."

"They get their foot in the door, and then it expands and expands and expands," one protester told the *Santa Fe New Mexican*. "I just feel we have to rein in some of this stuff. Let it stay up in Los Alamos."

Later, at an intersection near a different site, which the lab ended up leasing, opponents put up a sign that read,

YOU ARE LEAVING NEW MEXICO.
WELCOME TO LOS ALAMOS.

Regular Friday demonstrations have happened near there, featuring members of Veterans for Peace and Concerned Citizens for Nuclear Safety. "I'm sure people working in those buildings are very nice people," one demonstrator, Ken Mayers, told the *Albuquerque Journal*, "but the fact remains they are engaged in an evil enterprise."

Santa Fe's own paper, though, seems to disagree with the sentiment. "These are dollars Santa Fe desperately needs right about now," read an editorial from the publication's staff. "Welcome, Los Alamos National Laboratory Employees. We're glad to have you."

Tess Light is familiar with the animosity from the outside. It hit home hardest in 2000, Light recalls, when her world almost went up in flames. That summer, the Cerro Grande fire ripped through the town of Los Alamos, torching some four hundred houses and forty-three thousand acres, along with some of the lab's infrastructure. Light and her family were evacuated.

Down the hill, flames licking across the place she'd come to call home, Light went to a Target because she'd forgotten her toothbrush. The news was playing on a TV in the store, and the camera showed the park that was right across the street from her house: gone. "And I thought, 'My house is gone,'" she says. "And I start crying."

Two women in front of her were watching the news. "Isn't it terrible what's happening in Los Alamos?" one of them said.

"Yeah," said the other, "but they kind of deserve it."

Light, crying behind them, said nothing.

"I know we are the Antichrist," she says now, "and yet I think it's a bit more nuanced than that."

# CHAPTER NINETEEN

This is an exemplar of what can appear to
be deterrence (cars, don't hit me!) and is
compellence (allow me to make use of the
road or we both have a bad day!).

—Martin Pfeiffer

THE ANIMOSITY TOWARD WEAPONS SCIENTISTS STEMS AT LEAST IN PART
from not wanting to live under a mushroom-shaped shadow. A full 49
percent of Americans, according to a 2019 YouGov poll, want the United
States to work with other countries to completely eliminate nuclear
weapons.

On the other hand, if you ask the Mitchell Institute's 2021 survey, 91
percent of Americans agree with the statement "America's nuclear deterrence capability is critical to our national safety and security. It should be
one of the highest priorities of the Department of Defense."

Perhaps these things just tell us that polls are unreliable. Or that the
phrasing of questions influences answers. Regardless, opinions on the
value of nuclear weapons and deterrence, and what both of those mean
for us, are certainly varied. Most of those opinions are probably influenced by the lack of easily accessible information about the bomb and

bomb policy. There's little public debate or discussion about the role of nuclear weapons broadly, what other options exist, or how, specifically, the nuclear complex is moving into the future. And, in the United States, there is little understanding of how much preparation for that future is already taking place—and how close it comes to home.

Nevertheless, informed or not, consciously or not, all of us do live within that mushroom shadow. And, luckily for us, there are some scientists—social scientists—whose job is to understand the multifaceted nature of our relationship to the nuclear past and present. They're a small crew, these nuclear anthropologists, and one of their most prominent researchers is Hugh Gusterson of the University of British Columbia.

Gusterson's trajectory into nuclear study began in the early 1980s, when he read about a New York City protest against the nuclear arms race during a United Nations summit. Such protests were not so uncommon back then, at a time when a not insignificant (or at least not quiet) portion of the US population mobilized against the nuclear complex. Small towns passed resolutions declaring themselves to be "nuclear-free zones." "There was this sense, especially among young people, of global emergency," says Gusterson. The tone of the discussion was similar in octave to Gen Z's passion about climate change several decades later.

After this initial experience, Gusterson went on to work in activism. "I worked on the staff of the San Francisco Nuclear Freeze Campaign, and a teacher at a San Francisco high school asked if we'd send someone to debate a Livermore weapons designer in her class," he says. "That ended up being me."

"The part of me that was an anthropologist was quite fascinated to understand where he was coming from," he says, of the experience. "It was clear he understood he was right, and I was wrong."

But Gusterson, despite his activism, wasn't quite as straightforwardly convinced of his own views *himself.* Given his interests and his intellectual curiosity, perhaps it's no surprise that he did go on to pursue a doctorate in anthropology and chose fieldwork that was unconventional for the time. Most anthropologists focused—often focus still—on poorer people who live abroad. But Gusterson wanted to turn his mind toward US citizens with relatively high incomes: weapons workers at Lawrence

Livermore National Laboratory, people similar to the man he'd debated years prior. He traveled there to do his studies among the nuclear natives.

After his time at the California lab, Gusterson went on to write the book *Nuclear Rites*, published in 1996, in part about how nuclear scientists interact with the communities around them, how they work, and how they conceive of their work. In that research, Gusterson found an interesting way that nuclear scientists are perhaps different from the rest of us: when the researchers talked about weapons, they spoke in metaphors of birth, not—as might make more sense—death and destruction. It's a tradition that started early. When the Trinity explosion went off as planned in 1945, for example, the code words to indicate its success were "It's a boy."

Perhaps, in the scientists' worldview, birth was actually the correct parallel to draw. Many researchers do think of the Deterrent as savior, not destroyer. Nuclear weapons keep us from conflict, an attitude common within the national labs. Which makes sense, given another one of Gusterson's findings: that nuclear weapons scientists came from across the political spectrum and were able to think about their work in a way that gave it meaning and made it feel special. "What unified them was a faith that human beings are rational enough to control nuclear weapons, and not become their victims."

Historian Spencer Weart came to a similar conclusion. "Belief in the virtues of science and technology could be so strong that even a threat of destruction might sound like a promise of peace," he wrote in the 1988 book *Nuclear Fear*.

Weart argued that the birth analogy can extend even further. Let's say the bombs do go off. The world and its people burn. Humans, in whatever fractured form they remain, have to restart civilization, in whatever form it can take. They get a new life on a changed Earth. The detonations are, then, a kind of cleansing, sterilizing event. And humans, their lesson learned, can begin again, having paid for their prior sins. Bombs as re-creator, life giver, life keeper, savior. This line of thinking philosophically fuels the birth metaphor.

The bombs, though, aren't just like parts of the Holy Trinity. They are straight up Old Testament, moody and vengeful and beyond human control. In the Cold War, that led some people to a constant fear of nuclear apocalypse, scarred by schools' air-raid drills or fooled into thinking their expensive fallout shelters would keep them alive on this planet. But, eventually, the constant news of impending but never-arriving nuclear bombings jaded much of the population. Soon Americans simply embraced a kind of nihilism: they weren't in charge of the arsenals, couldn't stop nuclear detonations. And so they continued about their lives, giving bombs only passing thought.

In all these often contradictory expert and layperson views, the bombs have one thing in common: they are a kind of stand-in for gods. They create, save, destroy. They are out of mortal control, in charge of ends and beginnings, able to send you straight to heaven or leave you standing in hell. Plus, when a god is in charge, humans don't bear as much responsibility for the course of history.

And maybe, subconsciously or just poetically, we think about nuclear weapons this way because we sort of believe that we deserve that kind of punishment. Maybe even to atone for building the bombs in the first place.

Western culture, after all, is full of stories about people who were knocked flat for acquiring forbidden knowledge.

Adam and Eve, for instance, had to come live in our world.

When Gusterson published *Nuclear Rites* in 1996, weapons scientists were contending with the end of the Cold War and nuclear testing and the consequent start of stockpile stewardship. Destruction, creation, again, as always (never).

Today, though the national labs have been engaged in such nuclear nursing for decades, they largely haven't been doing the weapons design, production, and testing they did in Gusterson's day. Nuclear scientists and engineers had gotten used to their more passive pursuit, but Gusterson believes today's scientists might be sensing that skew. "The two key shifts that would signal to the weapons scientists that they're engaged in a different project is if they restarted testing or engaged in pit production," he says.

Some people fear the spool-up of the former, while the latter is straight up happening. The shifts herald not just a new nuclear era but a cultural change at National Nuclear Security Administration facilities, particularly at a place like Los Alamos. The New Mexico lab, Gusterson says, conceives of itself as an exploratory intellectual facility. But with pits on the docket, it takes on some of the character of a factory, meant to competently complete a known job, over and over, like the Savannah River Site.

But that possibility has always been there, underneath the lab's (and the labs') more creative work: the scientists' intellectual freedom was, at least in part, the will of the government that ruled their activities, their experimental pursuits existing only at federal pleasure.

When that will changes, so does the mission.

SINCE GUSTERSON CONDUCTED HIS WORK, OTHER ANTHROPOLOGISTS have followed in his footsteps, doing their own culture-centric studies on the nuclear complex. One scholar, Martin Pfeiffer of the University of New Mexico, focuses on how nuclear issues get communicated to the rest of us.

On Twitter, where Pfeiffer is best known, he goes by Martin "Doomsday" Pfeiffer or @NuclearAnthro. For a long while, a political cartoon pinned to his profile showed a mushroom cloud consuming and illuminating a city. "Let your faith light the world," it read. In the accompanying tweet, Pfeiffer wrote, "Nuclear weapons prevent war. Nuclear deterrence will work forever. There will be no accidents."

Clearly, he was being ironic. And one need only look at one of his catchphrases, of which he has several, to be sure. "We built them," Pfeiffer often writes. "We can take them apart. We can #LickTheBomb." These posts sometimes show a picture of Pfeiffer himself licking a bomb in a museum.

Pfeiffer would like to #LickTheBomb because he feels the world's nuclear arsenals are an existential threat to the planet. And that idea dogs him more than it does the average person. To wit: at some point on many days, he declares via social media using caps lock, "CLEANSING THERMONUCLEAR FIRE IS CANCELED FOR THE AFTERNOON!"

(or evening), implying—jokingly, of course—that such cleansing had been scheduled to take place. Usually, the occasion for the cancellation is that someone found a cute picture of a cat. Because internet.

And Pfeiffer is *on* the internet. He's not your typical academic. He's irreverent and speaks publicly and emotionally about his existential dread. ("My therapist has a really more detailed understanding of nuclear weapons than I think he ever wanted," Pfeiffer admits.) But when he's not doing that, he's trying to study how humans make meaning out of nuclear bombs, particularly how they do so at nuclear museums and heritage sites, like the splat of dirt where the Trinity test took place or the National Museum of Nuclear Science and History in Albuquerque, the city where Pfeiffer lives.

Pfeiffer noted the nuclear early, as a kid in the 1980s, when nuclear weapons made regular appearances on the news and in conversation. They received, he says, "a lot more attention than they do day-to-day today." One of his parents worked in nuclear power. One of his earliest memories is of the end of the Cold War, when the Berlin Wall came down, a razing that he watched on his grandparents' TV set.

Given that immersion, he doesn't remember when, specifically, he first heard of The Bomb—it was just part of life. But the conditions at home led him to a different sort of interest in it than his peers. Growing up in a physically abusive household, he felt powerless. Nuclear weapons, even if he himself didn't have access to them, represented might that could overpower even those who overpowered him. In the United States, he notes, there's a widespread tendency to look for decisively violent solutions to all problems. "Perhaps there's an argument to be made about it being an effort to find an utterly decisive form of violence," he muses.

It's definitely the musing of an anthropologist.

As an adult, figuring out what he wanted to study, he drifted back to that decisive form of violence. Nuclear weapons, he thought, deserved an anthropologist's focus. Researchers in anthropology are tasked with attempting to understand what humanity is, in all its forms, and how it came to be. For Pfeiffer, the possibility of humanity's potential end is part of that. Understanding how humans understand and interact with those weapons could, Pfeiffer believes, help upend the systems that make nuclear weapons seemingly necessary. The thought has similarities to how Paris feels about digging into the weapons codes. If we could just grasp it all, the thinking goes, maybe we could figure out we didn't need it.

Toward that end, Pfeiffer, who's pursuing his PhD, currently spends a lot of time wielding the Freedom of Information Act (FOIA) to get information out of the nuclear complex. He also takes lots of field trips to nuclear museums to see what their exhibits say (look at this tactical weapon)—and what they don't (that tactical weapon would have killed the person who deployed it). "These sites are not cookie-cutter," he says. They're sometimes influenced by local concerns, preoccupied, for instance, with the history or geology of the region. And they're sometimes influenced by factors like who sponsors an exhibit—as larger corporations or firms may set the agenda. There's the one about nuclear power sponsored by the nuclear-power company Urenco at the Albuquerque museum. And there's also the entry gallery to the National Atomic Testing Museum, sponsored by Lockheed Martin, which once ran the Nevada Test Site.

That's part of why Pfeiffer files so many FOIA requests too—to get the real story and synthesize the public and private agendas. It didn't work right away. His first FOIA request bounced right back to him. "I didn't have a clue what I was doing," he says.

Since then, Pfeiffer has gotten better at the task, and one of his favorite successes was the release of almost fifty years of Sandia's unclassified but not publicly archived employee newsletter, *LabNews*. It gives a window into how the lab saw itself, how it wanted employees to see things, and how it fit into the broader culture. In 1951, the earliest year in the archive, the lab used the word *girl* a lot and seemed preoccupied with security lapses.

### A Word of Warning

It has been brought to the attention of Sandia Corporation officials that there have been violations of security by employees of the Corporation and members of their families. Classified information concerning tests and other Corporation operations has been passed by "Q" cleared employees to members of their families and has been further disseminated by them. The

serious consequences these careless disclosures might have on the Atomic
Energy Program cannot be over-emphasized. The Espionage Act and the
Internal Security Act provide penalties for disclosing this type of information ranging from 10 years in prison and a $1,000 fine to 30 years in prison
or punishment by death.

Just a page later, another article implores employees to beware of
sneaky surveillance agents.

Have you seen any spies lately? Have you seen anyone slinking around the
tech area wearing a cape and a black hat with the rim turned down? No,
obviously not. That comic book routine was abandoned long ago by the
undercover chaps who make it their business to find out your business.
And because espionage and sabotage and all forms of subversive activities are such a refined subtle art, you may be lulled into the complacent
belief that we here at Sandia are immune to the treachery of information agents. . . . You were expecting maybe a hammer and sickle tattooed
on his forehead? . . . Don't go shooting off your mouth to others—even
though they are "Q-cleared." . . . The walls have ears.

Below that warning is, naturally, news of marriages and engagements.
Concern shifts over time, and world events have changed the lab's role
and focus. A 1992 newsletter, for example, took up the torch of nuclear
testing, reporting on the then secretary of energy's testimony against the
nuclear-testing moratorium. "Actual testing is an essential part of assurance that they remain safe while fulfilling their deterrent role," he said.

Nevertheless, one will note, testing is no longer an essential part of
assurance. The labs, today, publicly support a *lack* of testing. The truth,
enigmatic, has flipped in just a few years.

ANOTHER OF PFEIFFER'S FOIA WINS TOOK A MORE COLORFUL FORM. HE
received images of plutonium tetrafluoride (PuF4)—a form of plutonium that's part of the weapons-production process—from the complex.
In some of the pictures, the substance is a gray powder inside Petri-like

dishes and metal containers that look like part of a backpacking cook-ware set with the ashes of a campfire. In others, the PuF4 is in shards, or it's orange like dry clay, or it resembles ground-up rust. Sometimes it's fish food or cumin. Mineral makeup.

"DO NOT LICK ALPHA EMITTERS," Pfeiffer warned when he posted pictures on Twitter.

"Forbidden eyeshadow palette," someone commented.

But getting even information like that—pictures of an intermediary chemical or unclassified newsletters—hasn't been simple. Los Alamos first released only a single picture, initially, suggesting that that was responsive to his request because this was a representative image of PuF4. He appealed, noting that a world-class center for plutonium research since the 1940s would almost certainly have more than one such image. Magically, the organization found more.

From experience, he knows that whenever he files a FOIA request, he might have to fight for the information to be free. "How are you screwing me this time?" he says. "It's almost how it gets."

But he keeps going because, in general, he believes people should know more about what goes on within the nuclear complex and have the tools to more deeply examine what they know or think they know. That can all seem out of reach, since nuclear weapons are largely kept secret, despite efforts to bring sunshine to certain aspects of them, and because they are complex devices with highly technical explanations. But that hard science isn't the kind of knowledge you need to form your own ideas about nukes. "You're not necessarily required to know how to construct a thermonuclear weapon to be able to argue, 'Do we need more?'" says Pfeiffer.

That's part of why he posts his museum photos and documents online, in what he has humbly titled the Pfeiffer Nuclear Weapon and National Security Archive. It's also why he runs a Patreon site where he curates and analyzes his finds and shares and contextualizes his work in public: so people aren't bombarded with only raw, context-less information.

Giving people paper after paper isn't necessarily the way to go about things. "More information does not guarantee accountability or governance," he says. On the other hand, it's a necessary step in that direction.

"It's impossible to have accountability and governance *without* certain types of information," he says.

Perhaps it's unsurprising, at this point, that Pfeiffer doesn't exactly buy the idea of deterrence—as a military strategy or as a real concept. "There is no deterrence. There is only compellence" is another of Pfeiffer's online catchphrases. He sometimes illustrates the idea as a meme: a still from the 1984 *Ghostbusters* movie, in which Bill Murray is attempting to reason with Sigourney Weaver's character, who's been possessed by a spirit named Zuul.

"There is no Dana," she tells him in the film. "There is only Zuul."

"There is no deterrence," Pfeiffer writes in white text on the image. "Only compellence."

Besides, as Pfeiffer points out regularly, "Nukes aren't magic." They might matter less than we think. Maybe Russia doesn't *want* to attack us, so maybe we don't need to try to prevent them from doing so as fervently. "And that, of course, is the big unproven negative of deterrence," he says.

Perhaps without nuclear deterrence, he says, we could live in a world where people mutually care about each other and each other's humanity and countries don't assume worst-case scenarios about each other. Maybe something else, besides bombs, could bestow status on nations. "There are other ways toward power," Pfeiffer notes.

Maybe humans will get better at making decisions to save themselves long term.

Or maybe not. "And we will all burn one way or another," Pfeiffer says.

# CHAPTER TWENTY

Imagine a future where stability is not
precariously balanced on the threat of
mass destruction; where safety for some no
longer requires vulnerability for others.

—Global Zero

ANNE I. HARRINGTON ISN'T QUITE AS IRREVERENT AS MARTIN
Pfeiffer, and her ideas for reform come from within the system. Still,
she doesn't believe the official way of thinking, or talking, about nuclear
weapons is working. Strategies and policy around the use of The Bomb
often rest on the likely incorrect assumption that world leaders will not
overreact, underreact, misinterpret, or feel feelings at the expense of ra-
tionality. That's an estimation, and as an estimation, it is both easier to
deal with than reality and also inaccurate—a simulation smoother than
the world actually is. "Maybe we should start thinking about that," says
Harrington, an international relations scholar at Cardiff University.

She compares the current nuclear thinking to tools that economists use
and a strange creature they use to make sense of the world. Called *homo
economicus*, this fictional, generic dude behaves rationally with money all
the time, makes only decisions that make logical sense, and tries above all

to maximize dollars while minimizing work. He is *utilitarian* to the core. Economists have traditionally used him as a stand-in for the everyman.

But anyone who has engaged in an after-hours passion project or driven to an out-of-the-way grocery store to save $0.25 on a soda while guzzling $1 of gas knows this *homo* doesn't represent how real people actually behave.

In the nuclear world, *homo economicus*'s cousin is called, or at least Harrington calls him, *homo atomicus*: the guy who keeps it rational while contemplating whether to cause or how to avoid massive casualties and environmental contamination. *Homo atomicus* came of age during the Cold War, when analysts like those at the RAND Corporation— sometimes called the "Wizards of Armageddon"—worked out the basic math of deterrence theory and nuclear war plans for the first time. It was a new and quantifiably horrifying sort of activity. "Nobody had ever killed 35 million people on a sheet of paper before," wrote Fred Kaplan in a book titled, appropriately, *The Wizards of Armageddon*. The word used to quantify that level of human catastrophe was, at least according to one RAND intellectual, *megadeath*, meaning one million deaths from a nuclear detonation. Fairly, some people call the wizards "megadeath intellectuals." But after the Cold War, radioactive megadeath moved largely to the back of the American mind. It was there, sure—lurking, quiet. But potential nuclear Armageddon was a reality most of us just grew up with, and so it rested between our neurons as inherited knowledge. And like most passed-down ideas, it often went unexamined.

Those born after the Cold War also inherited nuclear arsenals that exist inertly to do their deterrent job and are more idea than object. "The threat of using the thing, not actually using the thing, is what matters," says Harrington.

Both of those ideas seemed to come under challenge in February 2022, though, when Russia began its attack on Ukraine and nuclear weapons rocketed to the top of public concern. It seemed like nuclear weapons put on a sequined dress and stepped to center stage. US leaders pointed to the very real nuclear issues raised by an invasion of eastern Europe, and to Russia's and China's continued modernization of their own weapons, to explain their own hefty spending on the US program.

Unlike during times when the nuclear sword dangled higher, almost invisible, above people's heads, when it lowered—as Russia lobbed explicit nuclear threats into the air and missiles into nuclear research centers and shells into power plants—people began to pay attention to the shiny weapons bumping up against their temples.

You can have *homo atomicus* running the show in every nuclear-armed nation. You can adopt a policy of minimum deterrence, keeping on hand only the number of weapons you need to put a foe off. You can even adopt Joseph Martz's idea of a capability-based, or weaponless, deterrent, in which countries retain the knowledge and infrastructure to whip up weapons in short order but don't keep them on hand. But regardless of which scaled-down deterrent you favor, world power whirls around atomic destruction—or at least the threat of it.

SOME IDEALISTIC EXPERTS WOULD LIKE TO SCRAMBLE THAT WORLD order. There should be no nuclear weapons, they say, no capacity to make them, and no variable that represents them in the international equation of state. That's the philosophy behind what some call the "global zero" movement.

Going no-weapons isn't a totally radical idea, although it sounds like one. It is, in fact, the philosophy underlying the Nuclear Nonproliferation Treaty, an extremely standard document that 191 countries have agreed to. As parties to it, states that didn't yet have nuclear weapons in 1967 agreed not to acquire them, and already-nuclear states agreed to work toward disarmament and disposal of weapons as the years passed. The first part has largely happened; the second part has had a rougher go of it. Given that the nuclear modernization budget is in the hundreds of billions, it doesn't seem that American nuclear weapons are going anywhere anytime soon. The same commitment is playing out in other nuclear nations, like Russia and China.

Then, too, in 2020, enough signatories signed the Treaty on the Prohibition of Nuclear Weapons that it entered into force the next year. The agreement, at least theoretically, bans the development, testing, deployment, stockpiling, use, and threat of using nuclear weapons. When countries with nuclear weapons sign it, they agree to a verifiable and time-delimited elimination of the nuclear arsenals.

Just one problem: none of the actual nuclear states *have* signed it; nor have many of the allies that live under the umbrella of their nuclear protection.

It is essentially meaningless, like all the scrawny kids signing a paper outlawing bullying. But maybe it's the thought that counts.

You could pin the global zero movement's origins on scientists with misgivings about the Manhattan Project or the camped-out activist movements of the Cold War, but its modern, slightly realistic form traces back to a 2007 *Wall Street Journal* op-ed headlined, straightforwardly, "A World Free of Nuclear Weapons." It wasn't written by some longhair or an overly optimistic policy intern. It was, instead, penned by a quartet of former federal power brokers: George Shultz, once secretary of state; Henry Kissinger, another former secretary of state and national security advisor; William Perry, previously secretary of defense; and Sam Nunn, a past senator. After the article came out, people referred to its authors as the Four Horsemen of the Apocalypse.

The heavyweights laid out concrete ways the United States could inch toward freedom from the atom and help the rest of the world do so too. It could stop keeping warheads poised for launch to make accidental or un-authorized use less likely. It could ratify the Comprehensive Nuclear-Test-Ban Treaty, make sure weapons and high-grade radioactive material got the best security, and work on regional conflicts so no one involved in them felt the need to get a bigger gun. Eventually, we could disappear the weapons entirely.

A year after the Horsemen's op-ed came out, the Global Zero organiza-tion formed, dedicated to lowering the number of nuclear weapons down to naught and laying out a similar concrete path toward its desired endpoint.

Emma Claire Foley wasn't there for the group's beginning. And she didn't know at the time, when she was in college, that she'd go on to work in nuclear weapons. But she'd been thinking about them, obliquely, for a long time.

She first became aware of them when she was a kid, obsessed with a book called *Sadako and the Thousand Paper Cranes*. It's a fictional account of a real girl named Sadako Sasaki, who survived the bombing of Hiroshima. As an adolescent, Sasaki came down with leukemia, which some called the "atomic bomb disease." Sasaki began to make paper cranes with origami, aiming to construct a thousand of them. Legend held that if you did so, one of your wishes would come true. In the book, Sasaki wishes to live and fails to get to her set number. She dies.

In real life, she actually exceeded her numerical goal. "I will write peace on your wings and you will fly all over the world," she reportedly said.

She also died in real life.

"It's one of these books they make kids read that are just so sad," says Foley.

In high school in the 2000s, Foley's exposure to the bomb continued, in the half-joking debate strategy her teachers presented. "The thing you learn to do in debate class and in debates was to have everything lead back to nuclear war," she says. "That was the way to win the debate. That was a strategy." It was reductive, she thought, but also revelatory about the way atomic weapons work in the real world: they are the ultimate trump card.

In college and graduate school, she studied Russian language and literature, as well as Russian and Eurasian history and culture. After college, she spent time in Ukraine, realizing that every issue in the country had been affected by the Chernobyl disaster. Finally, when she was leaving school and looking for jobs, all of this came together: she saw a listing with Global Zero on Indeed.com.

"The thing that I really respected about Global Zero's approach was that it was very ambitious," she says. "I think we're still at a point where if you say you want to eliminate nuclear weapons, you're gonna get laughed out of a lot of rooms. And I have seen that happen."

But when she looked at the organization's strategy and goals, she wasn't laughing: she thought their plan was rigorous, practical, stepwise. "We are really reckoning with the degree to which nuclear weapons form the foundation of the world order," she says.

Reckoning, in part, by recognizing that it's not humanly possible to just poof them away or have the United States say, "Okay, fine, they're bad." The plan has to involve maintaining the existing power balance on the way to a world free of nuclear weapons. In the first step of that plan, the nuclear powers start strategic stability talks. Russia and the United States back off their stockpiles. All nuclear nations speak to each other and also agree to never use nuclear weapons first.

In phase two, the three big powers cut their stockpiles further, and other nuclear nations agree to adhere to the same limit. More talking follows.

Third, nuclear states ratify something called the Global Zero Accords, which the organization describes as "a binding international treaty that removes all nuclear weapons from military service within two years, and requires the complete destruction of nuclear warheads by 2045."

And, finally, in the last part, all nuclear weapons actually do get removed from service. After that, monitoring and verification will happen indefinitely, to make sure everyone is upholding their end of the bargain.

It made sense to Foley then, and it still does five years into her work at Global Zero—at a time when nuclear weapons and fears are even more present, not just for people who work at no-nukes organizations. "I think that there's a larger reckoning that's happening," she says. "We're really coming to grips with how stuck we are in the nuclear order in certain ways. But it's the only organization that I know of that is grappling with how change might happen from start to finish."

But what about the stability deterrence supposedly provides? What of the plot showing deaths per capita in major wars? "That particular graph, I've seen many times," she says.

You can take issue with the math if you want or debate whether death integers are the most useful metric of destruction. More to the point, you could take issue with the "what if" of a single nuclear event in the future. "It's hard for people to understand the scale of destruction we're constantly holding ourselves an hour away from," Foley says. Purposeful assaults, retaliatory strikes, accidents, miscalculations. "There's a very real risk of a nuclear weapon being used in response to false information, or misinterpreted intelligence, or something like that," she says. "And so you look at that graph, and you're like, 'Yes, World War II is very bad.'" But the use of nuclear weapons, historians generally agree, wasn't the mono-causal end of the war, and the global peacetime that followed isn't neces-sarily the monoconsequential result of it.

"The fact that the only solution we've come up with is to hold ourselves at the very real risk of destruction many, many times greater than what occurred in those wars—that's not a solution," Foley says. "That's simply preparation for a much bigger problem."

But the global zero movement doesn't have the same sociocultural oomph its ilk did in the past. Forty years ago, Foley points out, a million people marched on New York City to oppose the existence of nuclear weapons, marking a UN meeting on disarmament. "The vast parade and rally, organized by a coalition of peace groups, brought together pacifists and anarchists, children and Buddhist monks, Roman Catholic bishops and Communist Party leaders, university students and union members. There were delegations from Vermont and Montana, Bangladesh and Zambia, and from many other places. The smiling, hand-clapping line of marchers was more than three miles long, and the participants car-ried placards in dozens of languages," read a *New York Times* article at the time. "A little girl carried a sign saying, 'I Hate Nuclear War,' and one marcher had an inflated rubber whale with the legend 'Save the Humans.'"

It's hard to imagine people mobilizing that way now, Foley says—even when Russia dangles the nuclear sword above the world's head (and so, because deterrence, nations like the United States dangle theirs back). But maybe someday that will change.

"When I talk to people about nuclear weapons, I want them to see themselves as part of that movement—the realistic long-term and really urgent effort to prevent the total destruction of human society," Foley says. "Which I really care about."


ARGUMENTS LIKE THOSE GLOBAL ZERO MAKES FOR DISARMAMENT aren't always effective with the general public, though—at least not ac-cording to Harrington's research. Focusing on the apocalyptic dangers sometimes leads people into the loving arms of deterrence theory. Disarmament advocates say we have to ditch nuclear weapons because nuclear war would be so terrible. "The deterrence theorists say, 'Yes, nuclear war is so terrible, and that's why we need nuclear weapons,'" Harrington says.

Nonexperts can see that logic too and are comforted by the security it seems to provide.

What *does* work in disarmament arguments, though, is focusing on the financial cost of nuclear weapons and related infrastructure—right now. In 2021, for instance, nuclear weapons got around $43 billion; the next year, the figure shot up by 17 percent, to $50.9 billion. Those meganumbers make a person wonder what else a government could do with that money.

In fact, contemplating that is what Foley has been doing: Her most recent project reimagines how the government might use the funds allocated to upgrade the siloed nuclear missiles across the Midwest and Rocky Mountain states. The feds could create new jobs for people who worked on the missiles, investing the cash in new infrastructure for those states, which are typically more economically depressed than the coastal states that don't have missiles buried beneath residents' feet. They could admit that a region-spanning field of missiles probably doesn't add much to the submarines and planes that also house missiles but does increase the likelihood that people will be killed on American soil. "We have this part of the country that is almost a sacrifice that is waiting to happen," says Foley.

Part of the reason people don't revolt against that potential sacrifice is that the missiles are pretty well hidden and not talked about much. "There's a great deal of effort on the part of people who manage America's nuclear weapons not to bring them to public attention," she says. But it's also true that federal budgets aren't a zero-sum game that allows for shuffling one spreadsheet cell's allocation to another. Money taken from nuclear modernization probably wouldn't go to new jobs. It would probably just go to different offenses and defenses.

THOSE WHO WANT TO ELIMINATE NUCLEAR WEAPONS AREN'T THE ONLY ones thinking about how to make the world safer and more secure (and reliable?). There are also people like Brad Roberts, one of Lawrence Livermore National Laboratory's big-picture guys, a strategy man who directs the Center for Global Security Research.

In May 2022, Roberts is wearing a puffy vest over a button-down shirt with Earth Shoe–style sandals. Behind him, a screensaver that looks like waving neon hairs blows in a digital breeze. Stacks of papers are arranged on shelves with printed labels like "Russia," "Trump Years," and "US Nuclear Strategy."

The center he directs is supposed to bridge the policy and technical worlds. "So that the technical community has a better understanding of the problems the policy people are trying to solve," he says. "And the policy people have a better understanding of the realm of the possible." Together, perhaps, technology + policy = global security. Or something.

Around Livermore, you can find a lot of technical people and many instances of the quote "We're the innovation lab." "And we don't wait for guidance," Roberts says. That may have worked well during the Cold War, but today it means that sometimes scientists are creating technology or doing research and development that no one asked for. "Here's our solution," he says, mimicking the researchers. "Now, where's your problem?"

You can see this in nonproliferation research and in test-detection programs, which can sometimes seem like research for knowledge's sake, since there's no promised path toward using the technology in the real world. That's true of Tess Light and Josh Carmichael's project, as well as Sandia's global-security research and the National Nuclear Security Administration–funded university initiatives on nonproliferation. The researchers, if they have the right phone numbers, can ask those who actually make decisions whether they are interested in using their technologies operationally—but they have no power themselves to make that operation a reality. In that way, some of the labs' applied science is actually more like basic science. It doesn't know where it's going and can't see its own future.

But figuring out how technical people can actually help policy people is particularly important right now, in Roberts's mind, because he believes the United States is working on Deterrence Strategy 3.0. Deterrence 1.0

defined the Cold War era: one-on-one confrontation between the United States and Russia. Deterrence 2.0 happened between the early 1990s and 2014, when officials worried more about rogue states, as opposed to one-on-one or one-on-two great power challenges.

But after 2014, when Russia annexed Crimea and China began upping its naval activities, things started to change. Since then, both countries have continued to modernize and revamp their nuclear forces. "Russia and China don't talk about stewardship," Roberts says. "They're doing something different." Plus, according to him, both countries are conducting nuclear pseudo-tests—ones they say are consistent with the Comprehensive Nuclear-Test-Ban Treaty because they're undetectable. Experimental test-detection systems are the kinds of things that Livermore could, theoretically, help usher into the real world, turning them from a technical project into political power. The threat of detection stopping pseudo-tests could help with deterrence on its own.

But deterrence, broadly, is dealing with more challenges and complications than it did in the past. "I think we're going to find the nuclear debate gets much more heated, contentious and difficult," Roberts says, "because the real world is pushing out of our comfort zone."

But for Roberts, that boundary pushing doesn't exactly lead to a global zero mind-set. In fact, in *The Case for U.S. Nuclear Weapons in the 21st Century*, Roberts lays out why the world would not be safer if the United States reduced or removed the bomb's role in its strategy.

When asked about his own book, though, Roberts begs off and instead cites another one. It's standing up on the top of his desk, its cover in full view. Flags emblazoned with buttered bread are caught in a still-frame flutter. Titled *The Butter Battle Book*, it was written by noted nuclear scholar Dr. Seuss. It was given to Roberts by a young air force officer who came for one of the lab's training programs. The man, at the end, thanked Roberts for the mentorship and handed him the slim orange volume. The officer said that his son had greeted him at the door recently. "Daddy, Daddy," the son said. "I know you think Brad Roberts is a genius, but this is what you really need to know about the Cold War."

His mother had put the kid up to it, but the sentiment stands.

*The Butter Battle Book*'s plot centers on two groups of people, split according to which side of their bread they butter. To separate themselves and denote their differences, they have built a wall.

Then the groups begin fighting across the wall, developing ever-more-capable weapons in response to the other side's ever-more-capable weapons (slingshots, Triple-Sling Jiggers, Jigger-Rock Snatchems, etc.). At the end of the book, both sides have a powerful bomb called the Big-Boy Boomeroo.

Mysteriously, even with these new weapons, their differences persist. "They give up on the whole thing and figure peace is a better way to live," says Roberts, "and accept that people butter their bread on opposing sides."

That's not, however, how most people would probably summarize the ending. At the close of the book, both sides hold equally powerful weapons. "Who's gonna drop it? Will you? Or he?" asks a child of his grandfather, who has the weapon in his possession. "Be patient!" says the grandfather. "We'll see."

Today, in the nonfictional world, we're all still we'll-seeing.

Roberts spends a lot of time considering what to do next, like what stockpile stewardship should look like after this modernization cycle is over and—not to get ahead of himself—whether the nuclear complex can even complete its current stewardship plans on time. But the problem facing the nuclear arsenal, he says, isn't the complexity of the problems themselves, a lack of money, or old infrastructure.

"It's culture," Roberts says. "It's 'Oh, well, you know, there's a twenty-seven-step review process you have to go through to do that. And then when you've done it with me, then you need to do twenty-six steps with them.' And everything's 'Mother, may I?' Everyone has a chance to say no."

As for deterrence, the author of *The Case for U.S. Nuclear Weapons in the 21st Century* doesn't actually know what it should or will look like—or even how well it works. He cites, as he must, another book, this one titled *Savage Century*. It's about how the twentieth century was, in many ways, a downhill slide and the one we're in now may be poised to look the same. "Is the twenty-first century going to be more or less savage than the twentieth? And will nuclear weapons be part of the reason that it was more savage or less savage?" he asks.

Roberts has been "collecting data" on those questions for years without coming to a conclusion. "You can find data pointing in both directions," he says. Nukes, he continues, "will either have a pacifying effect or a catastrophic effect. We don't know." And he doesn't have a guess.

"Monday, Wednesday, Friday, I feel one way. Tuesday, Thursday, Saturday, another. Then Sunday," he says, "I drink."

He probably drinks because he too is scared of Armageddon, or something a little less bad but still not great. Because he has his doubts about the radioactive tinder on which global dynamics rest. "I'd love to live in a world free of nuclear weapons," he says. "When people take a poll, I'm happy to vote that way."

But, he continues, nuclear threats exist. And he believes they can only, currently, be countered by further nuclear threats. "So long as they remain, we need a deterrent that's safe, secure, and effective," he intones. "And that requires more than just having bombs in the basement. It requires the intellectual architecture, it requires the policy architecture, it requires the ability to employ them in militarily significant ways."

It requires a bunch of national labs.

In his mind, though, that view of deterrence isn't actually at odds with global zero dreams. "So long as disarmament remains a remote goal, we need deterrence," he says. Nothing else we've found works. "So long as deterrence remains dangerous and unreliable, we need to continue to make progress toward disarmament."

Roberts turns toward his desk and points to Dr. Seuss's work. "If only it were that easy," he says.

# CHAPTER TWENTY-ONE

You think man can destroy the planet? What intoxicating vanity.

—Michael Crichton

It's a warm day in Los Alamos when Mark Paris sits in the shade of an umbrella near his office, which, in fall 2022, is no longer new to him. It is, though, he notes, right under a communications dish. "I haven't been sick since I started working in that office," he says. "So, I don't know, maybe the radiation is not bad enough yet to kill me, but it's killing all the germs."

He looks around and smiles at his own mimicry of conspiratorial thinking. "This is a joke," he clarifies. "It is nonionizing radiation."

He's at a table at Hot Rocks Java Cafe, a place whose name suggests the 1990s came up with it. The coffee shop is in a publicly accessible part of the Los Alamos National Laboratory (LANL) campus—although the guarded gate you have to pass through might suggest otherwise and provide its own sort of deterrence for casual coffee seekers.

Paris's student, Kyle Beling of the University of New Mexico, has been working on Ernest Rutherford's scattering problem, in which a singularity improbably appears. The results of his work aren't quite finished. This may seem like a trivial problem with inscrutable details, their implications bouncing off each other confusingly and uncertainly, like the particles themselves. But the point is that, within weapons-simulation circles, Paris continues to see complacency with results as they are and often a lack of curiosity about underlying assumptions or alternative methods.

Paris clearly subscribes a bit to the "just asking questions" model of physics—poking holes whose significance he's not sure of, both to try to understand how to fill them and simply to demonstrate that they exist, even if no one is looking at them.

"There are two cultures in science," he asserts. "One is 'take what's known now and solve problems by going around them. Just make a fix.' And then there's another which is 'solve the problem.' So I think there's a lot of going around the problem."

That could mean that you're on the 112th floor of the code, which scientists have been working on in their inertial way for years. The structure seems stable, but the foundation, which you haven't checked on, is crumbling.

"It's not as bad as I thought, in some places," he says, "and a whole lot worse than I ever imagined in others." Here, "bad" means that the physics in the codes hasn't been revisited since the 1940s or 1950s. "The calculational approaches that you develop are always with an eye toward the resources you have at the time that you're thinking about it," he says. "And those have changed."

Indeed: the phone Paris is not allowed to have in his building has more computing power than the mainframes the lab had in the mid-twentieth century. "But I don't know if the basic view has changed," he says.

When asked how specific he can be about the iffy parts of the code, he says, "No specifics."

When pressed lightly, he says, "Weapons primaries."

The failure to notice the flaws in the foundation was maybe aided, for a time, he suggests, by the lab's official organization, which didn't recognize nuclear anything, specifically. The lab calls these groupings "pillars," and in the pillars are united LANL's interdisciplinary works. For a while, there were only three:

1. Development and discovery of new materials
2. Ways of perceiving and identifying complex social, biological, climate-based, and other trends and interactions (referred to as science-based "signatures")
3. Prediction of threats to our nation and anticipation of other challenges to human survival, including in area of cybersecurity

There was no "nuclear" in the facility's priorities. "You take the most identifiable part of the laboratory and you throw it away," Paris says, almost guffawing in the way of a fictional character. "I think we've been recovering from that era for a while."

Today, the pillars include "nuclear and particle futures."

"I have no idea what a nuclear future looks like," Paris says. "Or a particle future. I have no idea."

Los Alamos does have a definition, which says that this pillar "focuses on the research required to maintain Los Alamos as the premier laboratory in United States for 'all things nuclear.'"

In Paris's mind, that means continuing to ask his needling questions—and doing a better job of incorporating the information LANL scientists have gathered, from old tests and new lab experiments, into the simulations. "There's a lot of data," he says. "I'm not sure there's a whole lot of time spent interpreting the data." He and others in his division are trying to pick that up and understand the data pouring in from experiments using a fundamental-physics perspective, so they can incorporate the actual behavior of nature, and the why behind it, into the simulations. "Without that, we're just kind of spinning the wheels," he says, "and that seems disingenuous, wasteful, fraudulent—all those scary words that people get thrown in jail for."

But the wheels keep spinning. After all, they haven't fallen off yet.

"You get into a routine, and it's tough to break," Paris says.

He still believes this understanding is important to a better nuclear future, although he's pretty uncertain about the nuclear present these days. Like everyone, he's worried a weapon might again, maybe soon, be used in anger. But if models of moving particles are hard to get right—if he cannot see *their* futures—that's even more true of the large mammals lording over this planet. "The biggest mistake I see scientists making is to extrapolate their ability to predict in highly prescribed systems," he says. "They tend to think they can generalize that to more complex systems like human relations. To me, it's folly and nonsensical to attempt that."

Given that, he can't say how likely he thinks it is that someone will detonate a nuclear weapon. But he does believe he can predict something about what he himself will do in that case—at least if he survives and so do his career choices. "If that terrible eventuality should occur, agnostic the perpetrator, it will change my life," he says. "I don't know exactly how, but I know it will. I will do something."

"I will certainly have to rethink working," he continues, "at a place like this."

He looks around, at the tables full of coffee sippers, the highway full of cars that made it through the gates, the buildings full of people imperfectly working out the futures of various things, particulate or otherwise. "I think that eventuality, which hopefully never occurs, calls into question the foundation or the existence of the mission," he says. As well as, he continues, the purpose and history of the lab.

It's a somewhat faltering, circuitous statement. But it means that if a nuclear bomb goes off, deterrence has failed. Its philosophy, which is the philosophy of stockpile stewardship, has been proven false.

In that case, keeping the weapons because they would keep the peace didn't, in fact, keep the peace. It only kept the weapons.

THAT SAME DAY, JOSH CARMICHAEL AND TESS LIGHT COME TO HOT Rocks Java Cafe and eavesdrop on this end of Paris's conversation. Once he's dropped his bombs, he says hello to them and then goes back to his nonionizing office.

Carmichael and Light are, at this point, one year into their Integrated Nuclear Detonation Detection (iNDD) project and preparing for the consequent review. "We were joking that you write a proposal, and you get thirty minutes, twenty minutes, to justify it in an oral presentation," says Light, sitting at an outside table with lenses transitioned to the autumn photons. "But then after a year, they want five hours to prove that you should have had that money. And it's like, 'Wow, that seems a little backward, but OK.'"

Backward and stressful. Light feels like they stood up in front of the funding committee and said, "'We've got this song, we know exactly what we're gonna do,'" she says. "And then it's three months in, and you're like, 'Well, for fuck's sake, that's not gonna work.' I'm feeling the pressure, man."

In case it's not clear, the project hasn't gone, in the months since March, quite as smoothly as the team had hoped. There's not a nuclear beta version they could have used in the event of a Russia-Ukraine eruption (which hasn't yet happened), because the regular beta version isn't quite working yet.

"We haven't gotten there because we've been collectively getting our shit together," explains Light.

One part of the shit is linguistic. People on the team speak different scientific dialects. The words they use can have two or more different meanings at the same time. As they have progressed, the unrealized superposition of those definitions has caused problems. Take the word *trigger*. "When I say my sensor triggered, I mean a specific thing," she says. "When Josh says it, he means a totally different thing. So you go through a month, and we're all talking about triggers and data, and then you realize we are talking about completely different things." Light means that a detector in space recorded an event that looks explosion-esque. Carmichael isn't talking about detectors at all; he is referring to algorithms spinning through their lines of code, calling "detection" or not.

Beyond the words, though, they've had to alter the method they were using. The data from the different sensors was just so different. Space-based instruments only send a trickle of information downlinking to Earth. Seismic sensors, on the other hand, record all the time and return petabytes. On top of that, space sensors record happenings at the speed of light, whether that light takes the form of radio waves, gamma rays, or regular old optical light. The ground-based sensors, on the other hand, catch signals that don't move quite as fast—at, say, the speed of sound.

Because of all that—disparate data rates and disparate timescales and just general disparateness—their original idea won't work. They can't simply add the data together into a cohesive stream all at once. "We literally had to do a complete pivot and figure out, 'Well, how the hell are we going to do this?'" says Light.

"Who knew science was such an anxiety-inducing exercise in humility?" she adds, and laughs.

But they got through that cortisol-filled destruction of ego and landed, maybe a little bruised, at the bottom with a solution. Instead of smashing all the data together at once, they have turned to a process they call *tip-and-cue*—a term usually used to describe how operators or algorithms tell satellites what to look at, based on information that comes in from a given region.

In iNDD's tip-and-cue, though, a satellite may see something that looks suspiciously bright going on below. That detection immediately triggers the software to look at the seismic data that's automatically collected all the time, then take a peek at the infrasound information and see if readings line up. And on and on to the other kinds of sensors. The ground-based data would be fused together next and then the combination fused with space data. "It's this almost tandem or serial set of checking for evidence of this thing, which may or may not be an explosion," says Carmichael.

There's only one real textbook on the specific sort of data fusion they're attempting, under inherent conditions of uncertainty. The textbook is in Carmichael's office, of course. And it was incomplete on how a system like this might work. So he wrote the author, who's still a professor and keeps up with the latest developments in the field, to ask if there had been new research that might help the detonation-detection team.

"Your work would be fundamentally new," the man said.

"So I thought, Well, if we weren't intimidated before, literally the guy that wrote the textbook on it was telling me 'I haven't done that yet,'" says Carmichael.

Problematically, "not writing new textbooks" was part of the pair's pitch to the lab—their promise that they were just doing something sensible that no one had attempted before, not trying to shake the scientific universe. Still, the harder work is, they believe, worth pursuing. "Nobody else is going to do it," says Light, shrugging.

Right now, they're testing the tip-and-cue method on simulations of explosions and data from real-world nonnuclear events. Specifically, ones they call *unicorns*, using the language of both threesomes and heavily funded startups. Carmichael defines these as "events that give you confusing signatures because they contain an unlikely set of scenarios superimposed together." For them, these are natural phenomena that should trigger lots of different sensors—like a bolide, or a large meteor, that happens to go fireball over a seismic and infrasound array near an air force base in July, when there's lots of background lightning data or a solar storm. "We're looking for these things that are confluences of everything, just all hell breaking loose naturally," says Light.

That once happened, for instance, above Thule Air Force Base in Greenland, the United States' northernmost outpost. "The signal was beautiful," says Carmichael, from Earth and from space. "You can see, 'Yes, something big and explosive happened right then.' And because it's over an air force base, naively it looks like a missile coming in, exploding over the base."

If that should happen again, they wouldn't want officials to get, in Carmichael's words, "nervous" and ready to retaliate. A fuser like theirs could help differentiate the two circumstances—a conventional shot from space versus a nuclear shot from another country.

Ideally, in these tests, sensor A might say, "Something happened, not sure what." Sensor B might echo that. But perhaps sensors C and D say, "Nada over here." Taken individually, A and B's pickups could be worrisome—each an imperfect indicator on its own. But putting them together with C and D's nondetections, scientists would know that A and B couldn't have seen a nuclear event. "The correct answer is to stand down, nothing to see here," says Light.

Soon(ish), when the team starts creating simulations of nuclear detonations and feeding synthesized data from those into their detection system, it should give them a more alarming answer. "Oh, no," says Light. "Assess this more closely."

THE SUN HAS BEGUN TO SLIDE DOWN THE SKY BY THE TIME LIGHT AND Carmichael have finished discussing their project's difficulties and the ways they failed to predict precisely how their future would go. How they thought everything was simpler than it was and how they felt sure they were the people who could easily figure it out.

At this point in the day, they have moved to contemplating their place at the lab, the lab's place in the world, and the place the world is in. It doesn't take long to get to this existential point, in conversations with people who spend their days anticipating and trying to inoculate against catastrophe.

As they venture into a philosophical realm, they riff off each other on the Hot Rocks patio. "I usually don't like to talk about politics at work, but I'm kind of an old-school liberal," says Carmichael. "People don't usually associate liberalism with patriotism—"

Light interrupts him. "Which pisses me off," she adds.

He nods. "I honestly feel like I'm serving my country working here at the lab," he says. What he's doing, he feels, matters. And that can be true whether or not anyone knows about his detection system or has heard the acronyms iNDD or CTBT and despite what they think of the existence of weapons labs. "What we're working on literally matters to people that make decisions about whether or not tomorrow is going to be nuclear Armageddon," he says.

If, that is, it works and makes the policy jump into operation.

Given those stakes, Light is often frustrated with the way people outside see her work. Every year, she says, people picket the lab, opposed to its mission, "The Mission." "I do recognize that this is a nuclear weapon laboratory," she says, looking up at the sky through her glasses. "And I do recognize that is a significant, dominant portion of the work that's done here." But, she continues, it's not the only thing. And yet people paint everyone who works at the lab with the same brush.

Recently she was reading a book by a climate scientist, about personal action and global warming. She enjoyed it and thought she might write him a note saying so. "And I get to the end of the book, and it says, you know, 'We each have to make our own choices. And so you can choose basically to leave the world a better place and to try and do this for your

children—or you can do something like work at the national labs.' Like
work at a weapon laboratory," she says. "And that was his specific example
of being evil."

Light, it should be clear, does not herself feel evil. She is not a weapons
scientist, after all. But the broad-brush issue isn't the only thing about
this characterization that irks her. It's that, despite the initial feelings she
herself had about the national labs, she doesn't actually think the people
who *do* work on weapons in the weapons laboratory are evil either.

"You know what? Nuclear weapons exist," she says, using her hands
to emphasize this uncomfortably irrefutable fact. "And I guess I would
rather have people who understand that rather than let them corrode
somewhere and potentially have a horrible accident."

Unless the country dismantled and did away with them, which neither
it nor any nuclear weapons state is headed toward doing in the near future, someone does need to accept their presence on this planet and keep
them from dangerous decay. "I was born into a world with them," she
says, fully exasperated at this point. "Not a fan, but that is the reality that
we're faced with. And I am comforted knowing that there's sincere and
very intelligent, hardworking people working to prevent those kinds of
things. So I think the national labs get a tough break."

If you're a critic, you could say that if she were *truly* not a fan, she
could actively work toward that dismantling idea and be more involved in
the global zero movement than the organization doing nuclear modernization. If you're a more moderate pragmatist—the kind of person, perhaps, who favors incremental rather than radical change, in general and
in arsenals—you might simply agree with Light. If you yourself are a fan,
you might perhaps think the argument is irrelevant: no need to disprove
the existence of evil on the mesas of Los Alamos, because places like Los
Alamos clearly help keep the world safe, and anyone who says otherwise
is simply naive.

Regardless of which opinion you hold, you could say that Light has
drunk the Kool-Aid or is rationalizing her participation in a destructive
enterprise. But you could equally say she took in new information and
changed her mind, and her life, and is working the best way she's found
to create the world as it ought to be.

All of this, in its own way, is probably a little bit true, and a little bit false, and a little bit unknowable. And in that, Light might agree with you. The world isn't a place where you can input neat variables and get an output with a clean conclusion, a seeable future, or even a present that makes sense. The data are messy, contradictory, confounding. They come in mismatched formats, are taken at inconvenient times.

"That's what I wish people maybe would appreciate," she says, "that the world is not black and white."

It's gray. Purple, blue, red, violet. Violent. Glowing. Thundering. Awesome. Terrifying. Prideful. Fearful. Regretful. Joyful. Anticipatory. Powerful. Waiting for its own unknown and unknowable futures.

# ACKNOWLEDGMENTS

I AM, FIRST, VERY GRATEFUL TO THE PEOPLE I INTERVIEWED, WHO shared their time, expertise, and insights. Their candor, thoughtfulness, intelligence, and willingness to have hard and probably annoying conversations are the only reasons this book exists. Thank you, too, to the public information officers who helped facilitate many of those interviews and to the security offices that cleared me, for whatever reason, to spend so much time at the national labs.

Thanks are due, too, to editors who published my related nuclear stories along the winding reporting path, helping sort out thorny topics, clarifying the thinking behind them, and ultimately clarifying the idea for *Countdown* itself. For this, I'm indebted to Katie Palmer, Clara Moskowitz, Michael Roston, Andrew Grant, Corinne Iozzio, Eric Hand, and Cynthia Hanson. Much gratitude to Sharon Weinberger for sharing her journalistic experience and guidance.

I'm always grateful for the support and enthusiasm of my agent, Zoe Sandler. And I'm very glad editor Remy Cawley at Bold Type Books was onboard with and helped shape the book in its early stages, and that Anu Roy-Chaudhury took that job on, too, and made the book better for the people who might eventually read its sentences. Paco Alvarez did a bang-up job with fact-checking, for which I am indebted.

I also owe many thanks to my family—my parents, Darla and Ron Scoles, and my sisters, Rachel and Rebekah. Spending time with them helped me keep perspective while I was thinking about scary potential futures. Brooke Napier is owed a thank-ye for many Montana porch discussions about deterrence, and also everything else. Ann Martin, my nuclear-adjacent companion, has been there and been smart and real, three ideal things.

Now, finally, whatever word is more than "thanks" belongs to my wife, Kim Stone, whom I met, somehow wooed, and married over the course of writing this book. She went to weird nuclear places with me, read bad drafts, talked atomic philosophy, and, most importantly, just existed from the beginning of every day to the end, a presence I will always be grateful to have around, and not just because I legally promised to.

This reporting was supported by the International Women's Media Foundation's Howard G. Buffett Fund for Women Journalists, which provided a grant that supported much of the travel required to do this topic some on-the-ground justice.

# REFERENCES

"Accelerated Strategic Computing Initiative (ASCI) Program Plan [FY2000]." DOE/ DP-99-000010592, Lawrence Livermore National Lab; Los Alamos National Lab; Sandia National Laboratories, January 1, 2000.

*Air Force Manual 91-221*. Department of the Air Force. March 26, 2020. https:// static.e-publishing.af.mil/production/1/af_se/publication/afman91-221/afman91 -221.pdf.

American Institute of Physics. "FY22 Budget Outlook: National Nuclear Security Administration." *FYI: Science Policy News*. August 20, 2021. https://www.aip.org /fyi/2022/fy22-budget-outlook-national-nuclear-security-administration.

Arms Control Association. "US Nuclear Excess: Understanding the Costs, Risks, and Alternatives." Arms Control Association. April 2019. https://www.usnuclear excess.org.

Associated Press. "Blog for LANL Workers to Shut Down." *Albuquerque Journal*. December 27, 2005. https://www.abqjournal.com/news/state/aplanlblog12-27-05.htm.

Atom Central. "Nuclear Weapon Effects on Trees." YouTube. https://www.youtube .com/watch?v=Q0kxCjiIyBA. Accessed May 18, 2022.

"Audit Report: DOE-OIG-20-33." Department of Energy. https://www.energy.gov/ig /downloads/audit-report-doe-oig-20-33 (accessed April 13, 2021).

Baker, Michele, and Christopher Silbernagel. *The Vega Subcritical Experiment (SCE) at the Nevada National Security Site U1a Complex*. DOE/NV/03624-0173. Nevada National Security Site/Mission Support and Test Services LLC. Office of Scientific and Technical Information. July 1, 2018. https://www.osti.gov/biblio/1471684-vega -subcritical-experiment-sce-nevada-national-security-site-u1a-complex.

Barigazzi, Jacopo. "Russia Defense Chief Makes Unfounded Claims of Kyiv Ready to Use 'Dirty Bomb.'" *POLITICO*. October 23, 2022. https://www.politico.eu /article/ukrainw-russia-vladimir-putin-volodymyr-zelenskyy-defense-chief-claims -of-kyiv-dirty-bomb.

Beyrer, Jack. "White House Gives In to Liberal Push to Stall Nuclear Modernization." *Washington Free Beacon*. August 12, 2021. https://freebeacon.com/national-security /white-house-stalls-nuclear-updates-in-hopes-of-appeasing-liberals.

Birch, Douglas. "Nuclear Weapons Lab Employee Fired After Publishing Scathing Critique of the Arms Race." Center for Public Integrity. July 31, 2014. https://

publicintegrity.org/national-security/nuclear-weapons-lab-employee-fired-after
-publishing-scathing-critique-of-the-arms-race.

Bloch, J., et al. The ALEXIS Mission Recovery. LA-UR-94-503. Office of Scientific and Technical Information. March 3, 1994. https://www.osti.gov/servlets/purl
/142518.

Blume, Lesley M. M. *Fallout*. New York: Simon & Schuster, 2020.

Boetel, Ryan. "Nuclear Weapons Spending to Get Boost in NM." *Albuquerque Journal*. June 2, 2022. https://www.abqjournal.com/2504641/nuclear-weapons-spending
-to-get-boost-in-nm.html.

Borgardt, James D., et al. "Results from the Third Galaxy Serpent Tabletop Exercise Demonstrating the Utility of Nuclear Forensics Libraries in Support of an Investigation." *Journal of Radioanalytical and Nuclear Chemistry* 322, no. 3 (December 2019): 1645–1656. doi:10.1007/s10967-019-06898-8.

Bozeman, Barry, and Larry Wilson. "Market-Based Management of Government Laboratories: The Evolution of the U.S. National Laboratories' Government-Owned, Contractor-Operated Management System." *Public Performance & Management Review* 28, no. 2 (2004): 167–185.

Broad, William J. *Star Warriors: A Penetrating Look into the Lives of the Young Scientists Behind Our Space Age Weaponry*. New York: Simon & Schuster, 1985.

"Bronco Thunder: NNSA and Interagency Partners Conduct Counterterrorism Exercise in Denver." Department of Energy. 2018. https://www.energy.gov/nnsa/articles
/bronco-thunder-nnsa-and-interagency-partners-conduct-counterterrorism.

Bunn, Matthew, Martin Malin, Nickolas Roth, and William Tobey. *Preventing Nuclear Terrorism: Continuous Improvement or Dangerous Decline?* Cambridge, MA: Belfer Center for Science and International Affairs, 2016.

Burns, Michael J. "LLNL/LANS Mission Committee Meeting." LA-UR-10-08082; LA-UR-10-8082, Los Alamos National Laboratory. December 6, 2010. Office of Scientific and Technical Information. doi:10.2172/1043557.

Caldicott, Helen. *The New Nuclear Danger: George W. Bush's Military-Industrial Complex*. New York: W. W. Norton, 2004.

Carver, M., and S. Tornga. "DIORAMA as a System Simulation and Modeling Tool." 2020 IEEE Aerospace Conference. Semantic Scholar. March 1, 2020. doi:10.1109/
AERO47225.2020.9172673.

Chenoweth, Erica, Paul Kolbe, Farah Pandith, and Nickolas Roth. "Domestic Terrorist Plots Against the U.S. Government: How Serious Is the Threat?" Belfer Center for Science and International Affairs. March 8, 2021. https://www.belfer
center.org/publication/domestic-terrorist-plots-against-us-government-how-serious
-threat.

Choi, Joseph. "US Discloses Size of Nuclear Stockpile for First Time Since 2018." *The Hill*. October 5, 2021. https://thehill.com/homenews/administration/575483
-us-discloses-size-of-nuclear-stockpile-for-first-time-since-2018.

Christian, Matthew. "Savannah River Site Pit Production Approvals Could Be Delayed by Six Months." *Post and Courier*. October 29, 2022. https://www.postand
courier.com/aikenstandard/news/savannah-river-site/savannah-river-site-pit
-production-approvals-could-be-delayed-by-six-months/article_32ac90f4-562e
-11ed-b75b-db2704f67186.html.

Cirincione, Joseph. "What Should the World Do with Its Nuclear Weapons?" *The Atlantic*. April 21, 2016. https://www.theatlantic.com/international/archive/2016/04/global-nuclear-proliferation/478854.

Clark, Carol. "NNSA Announces Successful Completion of First Production Unit of B61-12 Life Extension Program." *NNSA News*. December 3, 2021. https://ladailypost.com/nnsa-announces-successful-completion-of-first-production-unit-of-b61-12-life-extension-program.

Clark, David L., ed. *Plutonium Handbook*. 2nd ed. Downers Grove, IL: American Nuclear Society, 2019.

Clark, Douglas. "NNSA, Agencies Detail Radiological Incident Exercise." *Homeland Preparedness News*. May 16, 2022. https://homelandprepnews.com/stories/77033-nnsa-agencies-detail-radiological-incident-exercise.

Clements, Tom. "DOE Oversight Board Releases Critical Assessment of SRS Plutonium Pit Plant." SRS Watch. March 9, 2022. https://srswatch.org/doe-oversight-board-releases-critical-assessment-of-srs-plutonium-pit-plant-raises-safety-issues.

Clery, Daniel. "With Explosive New Result, Laser-Powered Fusion Effort Nears 'Ignition.'" *Science*. August 17, 2021. https://www.sciencemag.org/news/2021/08/explosive-new-result-laser-powered-fusion-effort-nears-ignition.

Cohn, Carol. "Sex and Death in the Rational World of Defense Intellectuals." *Signs: Journal of Women in Culture and Society* 12, no. 4 (July 1987): 687–718. doi:10.1086/494362.

Collins, Catherine, and Douglas Frantz. "The Long Shadow of A. Q. Khan." *Foreign Affairs*. August 2019. https://www.foreignaffairs.com/articles/north-korea/2018-01-31/long-shadow-aq-khan.

Committee on Radioactive Sources: Applications and Alternative Technologies. *Radioactive Sources: Applications and Alternative Technologies*. Washington, DC: National Academies Press, 2021. doi:10.17226/26121.

Communications Office. "Los Alamos Sensors Watch for Potential Nuclear Explosions." Los Alamos National Laboratory. January 31, 2018. https://www.lanl.gov/discover/science-briefs/2018/January/0131-vela-to-cubesats.php.

Congressional Budget Office. "Approaches for Managing the Costs of U.S. Nuclear Forces, 2017 to 2046." Congressional Budget Office. October 2017. https://www.cbo.gov/system/files/115th-congress-217-2018/reports/53211-nuclearforces.pdf.

———. "Projected Costs of US Nuclear Forces, 2021 to 2030." Congressional Budget Office. May 2021. https://www.cbo.gov/system/files/2021-05/57130-Nuclear-Forces.pdf.

Congressional Research Service. "The U.S. Nuclear Weapons Complex: Overview of Department of Energy Sites." Federation of American Scientists. March 31, 2021. https://fas.org/sgp/crs/nuke/R45306.pdf.

Considine, Laura. "Contests of Legitimacy and Value: The Treaty on the Prohibition of Nuclear Weapons and the Logic of Prohibition." *International Affairs* 95, no. 5 (September 2019): 1075–1092. doi:10.1093/ia/iiz103.

Cooper, Naomi. "NNSA Completes Exascale Computing Facility Modernization Project at LLNL." ExecutiveGov. June 9, 2022. https://executivegov.com/2022/06/nnsa-completes-exascale-computing-facility-modernization-project-at-llnl.

Cui, Yonggang. "Scoping Study of Machine Learning Techniques for Visualization and Analysis of Multi-source Data in Nuclear Safeguards." BNL-203606-2018-FORE, Brookhaven National Laboratory, Upton, NY. Office of Scientific and Technical Information. May 7, 2018. doi:10.2172/1436245.

Cunningham, James A., and Albert N. Link. "The Returns to Publicly Funded R&D: A Study of U.S. Federally Funded Research and Development Centers." *Annals of Science and Technology Policy* 6, no. 3 (March 2022): 228–314. doi:10.1561/110.00000023.

Defense Nuclear Facilities Safety Board. "Potential Energetic Chemical Reaction Events Involving Transuranic Waste at Los Alamos National Laboratory." DNFSB/TECH-46. Defense Nuclear Facilities Safety Board. September 2020. https://www.dnfsb.gov/documents/reports/technical-reports/potential-energetic -chemical-reaction-events-involving.

Demarest, Colin. "A Committee Kerfuffle over Plutonium Pits Foreshadows Bigger Battles." *Post and Courier.* September 11, 2021. https://www.postandcourier.com /aikenstandard/news/a-committee-kerfuffle-over-plutonium-pits-foreshadows -bigger-battles/article_7ed05794-1242-11ec-a699-eb31bb86fe03.html.

———. "Energy Department Suspends Search for Next Savannah River Site Management Team." *Aiken Standard.* November 10, 2021. https://news.yahoo.com /energy-department-suspends-search-next-005000305.html.

———. "The Nuclear Weapons Council Is Worried About Biden's Spending. So Are Activists." *Post and Courier.* July 31, 2021. https://www.postandcourier.com/aiken standard/news/the-nuclear-weapons-council-is-worried-about-bidens-spending-so -are-activists/article_eef5370c-f09a-11eb-b1eb-e708718a1640.html.

———. "Plutonium Pit Plan Involving SC Endorsed by Energy Secretary Jennifer Granholm." *Post and Courier.* May 6, 2021. https://www.postandcourier.com /aikenstandard/news/savannah-river-site/plutonium-pit-plan-involving-sc-endorsed -by-energy-secretary-jennifer-granholm/article_ce1126fc-aea2-11eb-8f43-b738 422c9d62.html.

———. "Rep. Smith Presses Biden to Audit Pit Production as Nuclear Posture Review Progresses." *Aiken Standard.* August 9, 2021. https://news.yahoo.com/rep-smith -presses-biden-audit-235300362.html.

Department of Defense. "Nuclear Posture Review." Department of Defense. April 2010. https://dod.defense.gov/Portals/1/features/defenseReviews/NPR/2010_Nuclear _Posture_Review_Report.pdf.

———. "Nuclear Posture Review." Department of Defense. February 2018. https:// media.defense.gov/2018/Feb/02/2001872886/-1/-1/1/2018-NUCLEAR -POSTURE-REVIEW-FINAL-REPORT.PDF.

———. "Nuclear Posture Review." Department of Defense. March 2022. https:// media.defense.gov/2022/Oct/27/2003103845/-1/-1/1/2022-NATIONAL -DEFENSE-STRATEGY-NPR-MDR.PDF.

Department of Energy. "Cleanup Sites." Department of Energy. https://www.energy .gov/em/cleanup-sites.

———. "DOE Announces Cybersecurity Programs for Enhancing Safety and Resilience of U.S. Energy Sector." Department of Energy. March 15, 2021. https:// www.energy.gov/articles/doe-announces-cybersecurity-programs-enhancing-safety -and-resilience-us-energy-sector.

———. "DOE Announces $18 Million to Advance Particle Accelerator Technologies and Workforce Training." Department of Energy. March 16, 2021. https://www.energy.gov/articles/doe-announces-18-million-advance-particle-accelerator-technologies-and-workforce-training.

———. "DOE Awards $17.3 Million for Student and Faculty Research Opportunities and to Foster Workforce Diversity." Department of Energy. May 3, 2021. https://www.energy.gov/articles/doe-awards-173-million-student-and-faculty-research-opportunities-and-foster-workforce.

———. "DOE Provides $28 Million to Advance Scientific Discovery Using Supercomputers." Department of Energy. July 16, 2021. https://www.energy.gov/articles/doe-provides-28-million-advance-scientific-discovery-using-supercomputers.

———. "Fiscal Year 2022 Stockpile Stewardship and Management Plan." Department of Energy. March 2022. http://lasg.org/documents/SSMP-FY2022_Mar2022.pdf.

———. *The Nuclear Matters Handbook: Expanded Edition*. Office of the Secretary of Defense. 2020. https://www.acq.osd.mil/ncbdp/nm/NMHB2020rev/docs/NMHB2020rev.pdf.

———. "Radioactive Gases Are Key to Detecting Nuclear Testing: NNSA and DOE Strengthen Security Through Sophisticated Sensing." Department of Energy. October 14, 2021. https://www.energy.gov/nnsa/articles/radioactive-gases-are-key-detecting-nuclear-testing-nnsa-and-doe-strengthen-security.

———. "Radiological Assistance Program, 60 Years." Department of Energy. October 2018. https://www.energy.gov/sites/prod/files/2018/10/f56/NNSA%20RAP%2060%20brochure%20-%20web.pdf.

———. "Spotlight: Quantum Information Science and Technology." Department of Energy. August 2020. https://www.energy.gov/sites/prod/files/2020/08/f77/OTT-Spotlight-Quantum-Information-Science-and-Technology.pdf.

———. "Women of Quantum Computing Go Tiny in Big Ways." Department of Energy. June 23, 2021. https://www.energy.gov/articles/women-quantum-computing-go-tiny-big-ways.

Department of Energy / Sandia National Laboratories. "Rare Open-Access Quantum Computer Now Operational." *EurekAlert!* March 15, 2021. https://www.eurekalert.org/pub_releases/2021-03/dnl-roq031521.php.

Doyle, James E. "Why Eliminate Nuclear Weapons?" *Survival* 55, no. 1 (January 2013): 7–34.

Easley, Mikayla. "Survey Finds Overwhelming Public Support for Spending on Nuclear Deterrence." *National Defense Magazine*. September 16, 2021. https://www.nationaldefensemagazine.org/articles/2021/9/16/survey-finds-overwhelming-public-support-for-spending-on-nuclear-deterrence.

Einstein, Albert. "Einstein's Letter to President Roosevelt—1939." *Atomic Archive*. August 2, 1939. https://www.atomicarchive.com/resources/documents/beginnings/einstein.html.

El Baradei, Mohamed. *The Age of Deception: Nuclear Diplomacy in Treacherous Times*. 1st ed. New York: Metropolitan Books/Henry Holt and Co., 2011.

Elkins, Kelly. "Nuclear Forensics: History, Selected Cases, Curriculum, Internship and Training Opportunities and Expert Witness Testimony." *Journal of Forensic Science*

*Education* 1, no. 1 (November 2019): https://jfse-ojs-tamu.tdl.org/jfse/index.php/jfse/article/view/2.

Ellis, Jason D., and Geoffrey D. Kiefer. *Combating Proliferation: Strategic Intelligence and Security Policy*. Baltimore: Johns Hopkins University Press, 2004.

Ellsberg, Daniel. *The Doomsday Machine: Confessions of a Nuclear War Planner*. London: Bloomsbury, 2017.

Eppich, Gary. "Origin Stories: How Nuclear Forensics Reveals the Histories of Smuggled and Trafficked Nuclear Materials." LLNL-JRNL-810544, National Nuclear Security Administration. Office of Scientific and Technical Information. May 20, 2020. https://www.osti.gov/servlets/purl/1756726.

Falcone, Michael. *The Rocket's Red Glare: Global Power and the Rise of American State Technology, 1940–1960*. Evanston, IL: Northwestern University, 2019.

Farmer, Annabella. "The War in Ukraine Has Put LANL's Nuclear-Weapons Mission in the Spotlight." *Searchlight New Mexico*. March 23, 2022. https://searchlightnm.org/the-ukraine-war-has-put-lanls-nuclear-weapons-mission-in-the-spotlight.

"50 U.S. Code § 2538a—Plutonium Pit Production Capacity." Legal Information Institute. https://www.law.cornell.edu/uscode/text/50/2538a.

*The Former Site of "LANL: The Real Story."* http://lanl-the-real-story.blogspot.com.

Frank, W. J., ed. "Summary of the Nth Country Experiment." Lawrence Radiation Laboratory, University of California, March 1967. National Security Archive. https://www.gwu.edu/~nsarchiv/news/20030701/nth-country.pdf.

Gainor, Chris. *The Bomb and America's Missile Age*. Baltimore: Johns Hopkins University Press, 2018.

Gallo, Marcy. "Federally Funded Research and Development Centers (FFRDCs): Background and Issues for Congress." Federation of American Scientists. December 1, 2017. https://fas.org/sgp/crs/misc/R44629.pdf.

Garberson, Jeff. "Deterrence Is No Longer Guaranteed." *The Independent*. June 2, 2022. https://www.independentnews.com/news/regional_and_ca/deterrence-is-no-longer-guaranteed/article_8844ff08-e291-11ec-a6b6-cbcea1932996.html.

Gardner, Timothy. "Advanced Nuclear Reactors No Safer Than Conventional Nuclear Plants, Says Science Group." *Reuters*. March 18, 2021. https://www.reuters.com/article/uk-usa-nuclearpower-idUSKBN2BA0CN.

*Gone Nuclear*. http://dontworrygonuclear.blogspot.com.

Government Accountability Office. "Department of Energy: Environmental Liability Continues to Grow, but Opportunities May Exist to Reduce Costs and Risks." GAO-21-585R. Government Accountability Office. June 8, 2021. https://www.gao.gov/assets/gao-21-585r.pdf.

———. "Preventing a Dirty Bomb: Vulnerabilities Persist in NRC's Controls for Purchases of High-Risk Radioactive Materials." GAO-22-103441. Government Accountability Office. July 2022. https://www.gao.gov/assets/gao-22-103441.pdf.

Gusterson, Hugh. "The Assault on Los Alamos National Laboratory: A Drama in Three Acts." *Bulletin of the Atomic Scientists* 67, no. 6 (November 2011): 9–18. doi:10.1177/0096340211426631.

———. *Nuclear Rites: A Weapons Laboratory at the End of the Cold War*. Berkeley: University of California Press, 1996.

Haddal, Risa. "NGSI International Nuclear Safeguards Engagement Program." SAND2015-0755PE, Sandia National Laboratories, Albuquerque, NM. Office of Scientific and Technical Information. February 1, 2015. https://www.osti.gov/biblio/1513948.

Hecht, Jeff. *Beam Weapons: Roots of Reagan's "Star Wars."* Auburndale, MA: Laser Light Press, 2015.

———. *Lasers, Death Rays, and the Long, Strange Quest for the Ultimate Weapon.* Buffalo, NY: Prometheus Books, 2019.

Hedden, Adrian. "Nuclear Weapons Development Coming Soon to Los Alamos National Laboratory amid Safety Concerns." *Carlsbad Current.* January 29, 2022. https://www.currentargus.com/story/news/local/2022/01/29/los-alamos-national-lab-prepares-nuclear-weapons-development/6562490001.

Higbie, P. R., and N. K. Blocker. "The Nuclear Detonation Detection System on the GPS Satellites." LA-UR-93-2834, Los Alamos National Laboratory. Office of Scientific and Technical Information. July 27, 1993. doi:10.2172/10185731.

Holian, Brad Lee. "Is There Really a Cowboy Culture of Arrogance at Los Alamos?" *Physics Today* 57, no. 12 (December 2004): 60–61. doi:10.1063/1.1878336.

Howieson, Susannah, Christopher T. Clavin, and Elaine M. Sedenberg. "Federal Security Laboratory Governance Panels: Observations and Recommendations." Institute for Defense Analyses. January 2013. https://apps.dtic.mil/sti/pdfs/ADA581271.pdf.

Hruby, Jill. "Senate Armed Services Committee Advance Policy Questions for Jill M. Hruby." Senate Armed Services Committee.

Hudson, Amy. "Modernizing the Triad." *Air Force Magazine.* March 23, 2022. https://www.airforcemag.com/article/modernizing-the-triad.

Hudson, Scott. "Radioactive Mercurial Alligators Once Roamed SRS." *Augusta Press.* September 6, 2021. https://theaugustapress.com/radioactive-mercurial-alligators-once-roamed-srs.

Hunter, David, Rhiannon T. Hutton, Matthew Breen, Patricia F. Bronson, William A. Chambers, Gregory A. Davis, Deena S. Disraelly, et al. "Independent Assessment of the Two-Site Pit Production Decision: Executive Summary." NS D-10711. Institute for Defense Analyses. May 2019. https://www.ida.org/-/media/feature/publications/i/in/independent-assessment-of-the-two-site-pit-production-decision-executive-summary/d-10711.ashx.

Ialenti, Vincent. "Drum Breach: Operational Temporalities, Error Politics and WIPP's Kitty Litter Nuclear Waste Accident." *Social Studies of Science* 51, no. 3 (June 2021): 364–391. doi:10.1177/0306312720986609.

Ichinose, G. A., S. R. Ford, K. Kroll, D. Dodge, M. Pyle, A. Pitarka, and W. R. Walter. "Preliminary Analysis of Source Physics Experiment Explosion-Triggered Microseismicity Using the Back-Projection Method." *Journal of Geophysical Research: Solid Earth* 126, no. 5 (2021): e2020JB021312. doi:10.1029/2020JB021312.

ICRC. "He Survived the Nuclear Explosion in Nagasaki, Then Spent His Life Treating Victims." *Humanitarian Law & Policy.* March 31, 2017. https://medium.com/law-and-policy/nagasaki-survivor-spent-life-treating-victims-4e6effdff9ee.

IIS Departmental. "Crisis in Ukraine: The Brink of Nuclear Catastrophe?" YouTube. https://www.youtube.com/watch?v=4Jv_c6-mDaI (accessed May 18, 2022).

International Physicians for the Prevention of Nuclear War International Commission and Institute for Energy and Environmental Research, ed. *Radioactive Heaven and Earth: The Health and Environmental Effects of Nuclear Weapons Testing In, On and Above the Earth: A Report of the IPPNW International Commission*. London: Zed Books, 1991.

James Martin Center for Nonproliferation Studies. "CNS Global Incidents and Trafficking Database." James Martin Center for Nonproliferation Studies. https://www.nti .org/analysis/resource-collections/the-cns-global-incidents-and-trafficking-database/.

Jervis, Robert, Richard Ned Lebow, and Janice Gross Stein. *Psychology and Deterrence*. Baltimore: Johns Hopkins University Press, 1985.

Jones, Joseph, Marty McRoberts, and Mary-Alena Martell. "Equipment Compatibility and Logistics Assessment for Containment Foam Deployment." SAND2005-5793, Sandia National Laboratories. Office of Scientific and Technical Information. September 1, 2005. doi:10.2172/876288.

Journal North Editorial Board. "LANL's Move to Santa Fe Means Jobs, and Controversy." *Albuquerque Journal*. May 9, 2021. https://www.abqjournal.com/2388532 /lanls-move-to-santa-fe-means-jobs-and-controversy-ex-the-reality-is-that-its -vast-amounts-of-money-and-jobs-make-lanl-with-an-annual-budget-north-of -3-billion-an-irresistible-force.html.

Kemp, R. Scott. "The Nonproliferation Emperor Has No Clothes: The Gas Centrifuge, Supply-Side Controls, and the Future of Nuclear Proliferation." *International Security* 38, no. 4 (April 2014): 39–78. doi:10.1162/ISEC_a_00159.

Kestenbaum, David. "Los Alamos National Lab Blog Draws Ire on Hill." NPR. May 19, 2005. https://www.npr.org/templates/story/story.php?storyId=4657337.

Kishner, Andrew. "U.S. Sneaks in 'Vega,' Its 28th Subcritical Nuclear Test." September 18, 2018. http://www.209days.com/vega.htm.

Knibbe, Morgan. "The Atomic Soldiers." *New York Times*. February 12, 2019. https:// www.nytimes.com/video/opinion/100000006186388/the-atomic-soldiers.html.

Kramer, David. "What Went Wrong with the Los Alamos Contract?" *Physics Today* 69, no. 3 (March 2016): 22–24. doi:10.1063/PT.3.3103.

Kristensen, Hans. "NNSA Nuclear Plan Shows More Weapons, Increasing Costs, Less Transparency." Federation of American Scientists. December 30, 2020. https:// fas.org/blogs/security/2020/12/nnsa-stockpile-plan-2020.

———. "Nukes in Europe: Secrecy Under Siege." Federation of American Scientists. June 13, 2021. https://fas.org/blogs/security/2013/06/secrecyundersiege.

Kristensen, Hans, and Matt Korda. "Estimating World Nuclear Forces: An Overview and Assessment of Sources." Stockholm International Peace Research Institute. June 14, 2021. https://www.sipri.org/commentary/topical-backgrounder/2021 /estimating-world-nuclear-forces-overview-and-assessment-sources.

*LANL: The Corporate Story*. http://lanl-the-corporate-story.blogspot.com.

*LANL: The Rest of the Story*. http://lanl-the-rest-of-the-story.blogspot.com.

LANL News. "Los Alamos National Laboratory Seeks Light Laboratory Space for Lease Within a 50-Mile Radius." *Los Alamos Reporter*. May 6, 2021. https:// losalamosreporter.com/2021/05/06/los-alamos-national-laboratory-seeks-light -laboratory-space-for-lease-within-a-50-mile-radius.

Larson, M. "Don't Mess with the NEST." Office of Scientific and Technical Information. March 15, 2012. https://www.osti.gov/servlets/purl/1047776.

Larzelere, Alex. "Delivering Insight: The History of the Accelerated Computing Initiative." Department of Energy. January 3, 2007. https://www.osti.gov/biblio/965460.

Last, T. S. "Big Development on the Nuclear Horizon." *Albuquerque Journal.* May 26, 2021. https://www.abqjournal.com/2394598/big-development-on-the-nuclear-horizon-ex-huge-growth-at-lanl-in-plutonium-pit-production-raises-ethical-issues.html.

Laub, Zachary. "The Impact of the Iran Nuclear Agreement." Council on Foreign Relations. April 11, 2017.

Lawrence Livermore National Laboratory. "From NIF to Z: LLNL Continues to Collaborate with Sandia on Technology Transfer Projects." Lawrence Livermore National Laboratory. June 8, 2021. https://www.llnl.gov/news/nif-z-llnl-continues-collaborate-sandia-technology-transfer-projects.

Leone, Dan. "Defense Nuclear Amendments for the Senate's FY22 NDAA." *Defense Daily.* November 16, 2021. https://www.defensedaily.com/defense-nuclear-amendments-for-the-senates-fy22-ndaa-w80-4-posture-review-plant-directed-r-more/congress.

———. "NNSA Can't Spend Its Way to Plutonium Pit Goal, STRATCOM Commander Tells Senators." *Defense Daily.* March 11, 2022. https://www.defensedaily.com/nnsa-cant-spend-its-way-to-plutonium-pit-goal-stratcom-commander-tells-senators/nuclear-modernization.

Lester, Paul. "What 'Stranger Things' Didn't Get Quite-So-Right About the Energy Department." Department of Energy. August 5, 2016. https://www.energy.gov/articles/what-stranger-things-didn-t-get-quite-so-right-about-energy-department.

Lewis, Jeffrey. "Minimum Deterrence." *Bulletin of the Atomic Scientists* 64, no. 3 (July 2008): 38–41. doi:10.2968/064003008.

Lewis, Nicholas, and Whitney Spivey. "Computing on the Mesa." Los Alamos National Laboratory. December 1, 2020. https://www.lanl.gov/discover/publications/national-security-science/2020-winter/computing-history.shtml.

Lewis, Peter D., and Jeffrey G. Zimmerman. "The Bomb in the Backyard." *Foreign Policy.* October 16, 2009. https://foreignpolicy.com/2009/10/16/the-bomb-in-the-backyard.

Lichterman, Andrew, and Jacqueline Cabasso. "Faustian Bargain 2000: Why Stockpile Stewardship Is Fundamentally Incompatible with the Process of Nuclear Disarmament." Western States Legal Foundation. May 2000.

Lilienthal, David E., Chester I. Barnard, J. R. Oppenheimer, Dr. Charles A. Thomas, and Harry A. Winne. *A Report on the International Control of Atomic Energy.* Washington, DC: Government Publishing Office, 1946.

Lin, Herbert. "Cyber Risk Across the U.S. Nuclear Enterprise." *Texas National Security Review.* June 21, 2021. https://tnsr.org/2021/06/cyber-risk-across-the-u-s-nuclear-enterprise.

———. "The Truth About Nuclear Deterrence." *IAI TV.* March 15, 2022. https://iai.tv/articles/the-truth-about-nuclear-deterrence-auid-2078.

Livermore Labs Events. "HEDS | Overview and Progress of Materials Experiments Using the NIF Ramp Compression Platform." YouTube. August 18, 2021. https://www.youtube.com/watch?v=fj9GlmbX8kQ&list=Ply9rIbGDXrG3noqQ4wkG6DoMACYoKiulr&index=19.

Los Alamos National Laboratory. "Foundation of Stockpile Confidence." Los Alamos National Laboratory. August 31, 2021. https://discover.lanl.gov/publications/the-vault/2021/foundation-of-stockpile-confidence.

———. "In Their Own Words: Alan Carr." 1663, Aug. 2018. https://cdn.lanl.gov/files/1663-32-august2018_2a1ce.pdf.

———. "Keepin Nonproliferation Program Grad Takes a Closer Look at Nuclear Forensics Chemistry." Los Alamos National Laboratory. January 2021. https://www.lanl.gov/discover/science-briefs/2021/January/0113-keepin-nonproliferation-program.php.

———. "Lab Director Thom Mason Hosts June 14 Public Meeting." YouTube. June 20, 2020. https://www.youtube.com/watch?v=sqqgeb1UM7s.

———. "MaRIE: A Facility for Time-Dependent Materials Science at the Mesoscale." Los Alamos National Laboratory. 2015. https://www.lanl.gov/science-innovation/science-facilities/dmmsc/_assets/docs/brochure.pdf.

———. "Nuclear Weapon Simulation and Computing." Advanced Simulation and Computing (ASC) Program. https://www.lanl.gov/projects/advanced-simulation-computing.

———. "Platforms." Advanced Simulation and Computing (ASC) Program. https://www.lanl.gov/projects/advanced-simulation-computing/platforms/index.php.

———. "Protecting Against Nuclear Threats." Los Alamos National Laboratory. 2022. https://mission.lanl.gov/nuclear-threats/.

———. "Science of Signatures: 2014 Update." Los Alamos National Laboratory. 2014. https://www.lanl.gov/science-innovation/_assets/docs/Science%20of%20Signatures2014.pdf.

———. "Science Pillars." Los Alamos National Laboratory. September 13, 2012. Internet Archive. https://web.archive.org/web/20120913040900/http://www.lanl.gov/science-innovation/pillars/index.php.

———. "Space and Remote Sensing (ISR-2)." Los Alamos National Laboratory. https://www.lanl.gov/org/ddste/aldgs/intelligence-space-research/space-remote-sensing/index.php.

———. "Unauthorized Drone Flights Prohibited in LANL Restricted Airspace, Including Additional No Drone Zone." *Los Alamos Reporter*. August 24, 2021. https://losalamosreporter.com/2021/08/23/unauthorized-drone-flights-prohibited-in-lanl-restricted-airspace-including-additional-no-drone-zone.

"Los Alamos National Laboratory Community Leaders Study." Los Alamos National Laboratory. September 2005. https://www.lanl.gov/community/_assets/docs/2005-community-leaders-survey.pdf.

Los Alamos Study Group. "Bulletin 287: Nov 5 Demonstration and Workshops; NM Greenhouse Gas Emissions Have Risen by About Half Under the Current Administration; Legislative Testimony on Plutonium Pit Production." Los Alamos Study Group. November 22, 2021. http://www.lasg.org/ActionAlerts/2021/Bulletin287.html.

———. "Please Come to the State Capitol on Friday, Nov. 5, Noon to 5 Pm: Update; New Full-Page Ads for Your Use; Help Us Get the Word out Please!" Los Alamos Study Group. October 24, 2021. http://lasg.org/letters/2021/nm_24Oct2021.html.

Los Angeles Air Force Base. "Space Systems Command Successfully Launches Space Test Program-3 Mission." SpaceRef. December 7, 2021. http://spaceref.com/news /viewpr.html?pid=58919.

Mahoney, Noi. "What's It Like Hauling Nuclear Weapons Across the Country?" *FreightWaves*. July 18, 2021. https://www.freightwaves.com/news/whats-it-like -hauling-nuclear-weapons-across-the-country.

Marciscano-Bettis, Raiza. "Groups Fire Back at Feds' Move to Dismiss Plutonium Pit Lawsuit." Tri-Valley CAREs. October 26, 2021. https://trivalleycares.org/2021 /groups-fire-back-at-feds-move-to-dismiss-plutonium-pit-lawsuit.

Martz, Joseph C. *Without Testing: Stockpile Stewardship in the Second Nuclear Age*. LA-UR—14-20080, 1114405. Office of Scientific and Technical Information. January 7, 2014. doi:10.2172/1114405.

Matasick, Marcy. "When Will New Mexicans Say Enough Is Enough to Los Alamos Lab?" *Santa Fe New Mexican*. August 8, 2021. https://www.santa fenewmexican.com/opinion/my_view/when-will-new-mexicans-say-enough -is-enough-to-los-alamos-lab/article_c0410c00-f7f8-11eb-9210-9bd245180752 .html.

Matlock, Staci. "A History of Innovation and Dysfunction at Los Alamos National Laboratory." *Santa Fe New Mexican*. January 2, 2016. https://www .santafenewmexican.com/news/local_news/a-history-of-innovation-and-dysfunction -at-los-alamos-national-laboratory/article_6bde4aee-077f-56a6-836b-eab1d289 271e.html.

Matthews, Patrick. "Corrective Action Investigation Plan for Corrective Action Unit 375: Area 30 Buggy Unit Craters, Nevada Test Site, Nevada." DOE/NV-1364. Navarro Nevada Environmental Services. Office of Scientific and Technical Information. March 1, 2010. doi:10.2172/975050.

Mattis, Jim. "Summary of the 2018 National Defense Strategy of the United States of America: Sharpening the American Military's Competitive Edge." Department of Defense. https://www.defense.gov/Portals/1/Documents/pubs/2018-National -Defense-Strategy-Summary.pdf.

McCrisken, Trevor, and Maxwell Downman. "'Peace Through Strength': Europe and NATO Deterrence Beyond the US Nuclear Posture Review." *International Affairs* 95, no. 2 (March 2019): 277–295. doi:10.1093/ia/iiz002.

McNamara, Laura. "Ways of Knowing About Weapons: The Cold War's End at the Los Alamos National Laboratory." PhD diss., University of New Mexico, Albuquerque, 2001.

Medalia, Jonathan. "U.S. Nuclear Weapon 'Pit' Production Options for Congress." R43406. Federation of American Scientists. February 21, 2014. https://sgp.fas.org /crs/nuke/R43406.pdf.

Mello, Greg. "Bulletin 296: The Troubled Logistics of LANL Pit Production: How Will LANL Staff and Contractors Get to Work?" Los Alamos Study Group. March 26, 2022. http://www.lasg.org/ActionAlerts/2022/Bulletin296.html.

———. "Can Santa Fe Survive as a Nuclear Weapons Suburb?" *Santa Fe New Mexican*. January 15, 2022. https://www.santafenewmexican.com/opinion/my_view /can-santa-fe-survive-as-a-nuclear-weapons-suburb/article_b6ab8ce8-7556-11ec -b47a-57273af4ebbc.html.

MITRE. "FFRDCs—a Primer: Federally Funded Research and Development Centers in the 21st Century." MITRE. April 2015. https://www.mitre.org/sites/default/files/publications/ffrdc-primer-april-2015.pdf.

Montgomery, Paul L. "Throngs Fill Manhattan to Protest Nuclear Weapons." *New York Times.* June 13, 1982. https://www.nytimes.com/1982/06/13/world/throngs-fill-manhattan-to-protest-nuclear-weapons.html.

Morland, Howard. "The H Bomb Secret: To Know How Is to Ask Why." *The Progressive.* November 1979.

Mount, Adam. "Trump's Troubling Nuclear Plan." *Foreign Affairs.* February 2018. https://www.foreignaffairs.com/articles/2018-02-02/trumps-troubling-nuclear-plan.

Myers, Steven Charles. *Radiation Signatures of Potential Nuclear Threats.* LA-UR-19-24078, Los Alamos National Laboratory. Office of Scientific and Technical Information. April 8, 2019. doi:10.2172/1511601.

Nakatani, Akio. *Death Object: Exploding the Nuclear Weapons Hoax.* Scotts Valley: CreateSpace Independent Publishing Platform, 2017.

Nanos, Peter. *Suspension of All Activities. Memorandum from Office of the Director to All Employees.* DIR-04-242. Los Alamos National Laboratory. 2004. http://www.lanl.gov/orgs/pa/newsbulletin/2004/07/19/DIR-04-242_Standdown.pdf.

NARAC. "Event Timeline." Lawrence Livermore National Laboratory. https://narac.llnl.gov/about/event-timeline.

Narayanamurti, Venkatesh, Laura Diaz Anadon, Gabe Chan, and Amitai Bin-Nun. "Securing America's Future: Realizing the Potential of the DOE National Laboratories." Belfer Center. October 28, 2015. https://www.belfercenter.org/sites/default/files/files/publication/testimony-narayanamurti-diazanadon-chan-bin-nun%20v2.pdf.

National Academies of Sciences, Engineering, and Medicine. *Governance and Management of the Nuclear Security Enterprise.* Washington, DC: National Academies Press, 2020.

———. *Nuclear Proliferation and Arms Control Monitoring, Detection, and Verification: A National Security Priority: Interim Report.* Washington, DC: National Academies Press, 2021. doi:10.17226/26088.

National Nuclear Security Administration. "FY 2022 Presidential Budget for NNSA Fully Funds Nation's Nuclear Security Enterprise." *Los Alamos Daily Post.* May 28, 2021. https://ladailypost.com/fy-2022-presidential-budget-for-nnsa-fully-funds-nations-nuclear-security-enterprise.

———. "Innovation at Our Fingertips: Making the Nuclear Security Enterprise SAFER." Department of Energy. May 12, 2021. https://www.energy.gov/nnsa/articles/innovation-our-fingertips-making-nuclear-security-enterprise-safer.

———. "International Nuclear Safeguards Engagement Program." Department of Energy. https://www.energy.gov/sites/prod/files/2020/07/f76/International%20Nuclear%20Safeguards%20Engagement%20Program%202018.pdf.

———. "National Nuclear Security Administration Strategic Vision." Department of Energy. May 2022. https://www.energy.gov/sites/default/files/2022-05/20220502%20NNSA%20Strategic%20Vision.pdf.

———. "NNSA Approves Critical Decision 1 for Savannah River Plutonium Processing Facility." Department of Energy. June 28, 2021. https://www.energy.gov/nnsa/articles/nnsa-approves-critical-decision-1-savannah-river-plutonium-processing-facility.

———. "Notice of Intent to Prepare a Site-Wide Environmental Impact Statement for Continued Operation of the Los Alamos National Laboratory." *Federal Register*. August 19, 2022. https://www.federalregister.gov/documents/2022/08/19/2022-17901/notice-of-intent-to-prepare-a-site-wide-environmental-impact-statement-for-continued-operation-of.

———. "President's FY-23 NNSA Budget Enables 'Responsive and Responsible' Nuclear Security Efforts." *Los Alamos Daily Post*. March 28, 2022. https://ladailypost.com/presidents-fy-23-nnsa-budget-enables-responsive-and-responsible-nuclear-security-efforts.

———. "Prevent, Counter, and Respond—NNSA's Plan to Reduce Global Nuclear Threats, FY 2020–FY 2024." Department of Energy. July 2019. https://www.energy.gov/sites/default/files/2019/07/f65/FY2020_NPCR.pdf.

———. "RadSecure 100: How NNSA Enhances Radiological Security Across the Nation." Department of Energy. September 2, 2021. https://www.energy.gov/nnsa/articles/radsecure-100-how-nnsa-enhances-radiological-security-across-nation.

———. "Request for Information Regarding Establishment of the Department of Energy Uranium Reserve Program." 6450-01-P. *Federal Register*. September 13, 2021.

———. "Stockpile Stewardship and Management Plan—Biennial Plan Summary." Department of Energy. December 2020. https://www.energy.gov/sites/prod/files/2020/12/f82/FY2021_SSMP.pdf.

———. "W87-1 Modification Program Fact Sheet." Department of Energy. January 2022. https://www.energy.gov/nnsa/articles/w87-1-modification-program.

"The National Nuclear Security Administration Has Completed the First Production Unit of a Modernized Warhead." MyHighPlains.com. July 14, 2021. https://www.myhighplains.com/news/local-news/the-national-nuclear-security-administration-has-completed-the-first-production-unit-of-a-modernized-warhead.

"National Quantum Initiative Act—H.R.6227/S.3143." FYI Science Policy Initiative from AIP. July 24, 2018. https://www.aip.org/fyi/federal-science-bill-tracker/115th/national-quantum-initiative-act.

National Research Council. *Aligning the Governance Structure of the NNSA Laboratories to Meet 21st Century National Security Challenges*. Washington, DC: National Academies Press, 2015. doi:10.17226/19326.

———. *Getting Up to Speed: The Future of Supercomputing*. Washington, DC: National Academies Press, 2005.

———. *Impact of Advances in Computing and Communications Technologies on Chemical Science and Technology: Report of a Workshop*. Washington, DC: National Academies Press, 1999. doi:10.17226/9591.

———. *Managing for High-Quality Science and Engineering at the NNSA National Security Laboratories*. Washington, DC: National Academies Press, 2013.

———. *The Quality of Science and Engineering at the NNSA National Security Laboratories*. Washington, DC: National Academies Press, 2013.

*National Security Science* Staff. "Pit Production Explained." Los Alamos National Laboratory. December 12, 2021. https://discover.lanl.gov/publications/national-security-science/2021-winter/pit-production-explained.

Nelson, Matthew. "DOE Unveils Funding Opportunities for Data Science, Computing Research Projects." ExecutiveGov. March 22, 2021. https://www.executivegov

.com/2021/03/doe-unveils-funding-opportunities-for-data-science-computing
-research-projects.

Nevada National Security Site. "Protecting the NNSS from Wildland Fires." Nevada National Security Site. https://www.nnss.gov/docs/fact_sheets/NNSS-WILD
-U-0043-Rev01.pdf.

———. Source Physics Experiments (SPE). https://www.nnss.gov/docs/fact_sheets
/NNSS-SPE-U-0034-Rev01.pdf.

"New Mexico Plans Los Alamos Community Meeting This Month." *Exchange-Monitor*. February 15, 2021. https://www.exchangemonitor.com/new-mexico
-plans-los-alamos-community-meeting-month.

Newdick, Thomas. "Newly Declassified Data Shows Unexplained Increase in U.S.
Nuclear Warhead Stockpile." The Drive. https://www.thedrive.com/the-war
-zone/42666/newly-declassified-data-shows-unexplained-increase-in-u-s-nuclear
-warhead-stockpile.

Niemeyer, S., and L. Koch. *The Historical Evolution of Nuclear Forensics: A Technical Viewpoint*. IAEA-CN-218-117. International Atomic Energy Agency. https://
www-pub.iaea.org/MTCD/Publications/PDF/SupplementaryMaterials/P1706
/Plenary_Session_1A.pdf.

"NNSA Elite Unit Celebrates 20th Anniversary." Department of Energy. https://www
.energy.gov/nnsa/articles/nnsa-elite-unit-celebrates-20th-anniversary.

Nuclear Engineering International. "US Funding for Accelerator Technology." *Nuclear Engineering International*. March 18, 2021. https://www.neimagazine.com/news
/newsus-funding-for-accelerator-technology-8609366.

Nuclear Threat Initiative. *Innovating Verification: New Tools & New Actors to Reduce Nuclear Risks: Overview*. Nuclear Threat Initiative. July 2014. https://media.nti.org/pdfs
/VPP_Overview_FINAL.pdf.

———. *Innovating Verification: New Tools & New Actors to Reduce Nuclear Risks: Verifying Baseline Declarations of Nuclear Warheads and Materials*. Nuclear Threat Initiative. July 2014. https://media.nti.org/pdfs/WG1_Verifying_Baseline
_Declarations_FINAL.pdf.

Office of Environment, Health, Safety, and Security. "Accident Investigation of the August 21, 2012, Contamination Incident at the Los Alamos Neutron Science Center at the Los Alamos National Laboratory." Department of Energy. September 18, 2012. https://www.energy.gov/ehss/downloads/accident
-investigation-august-21-2012-contamination-incident-los-alamos-neutron.

———. "Annual DOE Occupational Radiation Exposure | 2018 Report." Department of Energy. January 7, 2020. https://www.energy.gov/ehss/downloads/annual
-doe-occupational-radiation-exposure-2018-report.

"ORISE Report Shows Overall Number of Nuclear Engineering Degrees Increases to Highest Level since 2016." Oak Ridge Institute for Science and Education. February 2, 2021. https://orise.orau.gov/news/archive/2021/nuclear-engineering
-degrees-report-2019.html.

Panda, Ankit. "'No First Use' and Nuclear Weapons." Council on Foreign Relations.
July 17, 2018. https://www.cfr.org/backgrounder/no-first-use-and-nuclear-weapons.

Pardue, Doug. "Deadly Legacy: Savannah River Site near Aiken One of the Most Contaminated Places on Earth." *Post and Courier*. May 21, 2017. https://

www.postandcourier.com/news/deadly-legacy-savannah-river-site-near-aiken
-one-of-the-most-contaminated-places-on-earth/article_d325f494-12ff-11e7-9579
-6b0721ccae53.html.

Patterson, Eileen. "The Double Flash Meets the Bhangmeter." *National Security Science.* July 2015. https://cdn.lanl.gov/files/nss-july-2015_219c6.pdf.

Perkovich, George. "Reinventing Nuclear Arms Control." Carnegie Endowment for International Peace. September 9, 2020. https://carnegieendowment.org/2020/09/09/reinventing-nuclear-arms-control-pub-82500.

Perry, William James. *My Journey at the Nuclear Brink.* Stanford, CA: Stanford University Press, 2015.

Pincus, Walter. "A Deep Dive on the US' Nuclear Weapons Testing Tunnel." *The Cipher Brief.* July 20, 2021. https://www.thecipherbrief.com/column_article/a-deep-dive-on-the-us-nuclear-weapons-testing-tunnel.

Pongratz, Morris B. "Los Alamos in Space: The 50 Years Since Vela." LA-UR-13-27772, Los Alamos National Laboratory. Office of Scientific and Technical Information. October 7, 2013. doi:10.2172/1095885.

Priedhorsky, William. "Eyes in Space: Sensors for Treaty Verification and Basic Research." *Los Alamos Science* 28 (2003): https://fas.org/sgp/othergov/doe/lanl/pubs/las28/priedhorsky.pdf.

Quantum Information Science and Engineering Network. "Overview." Quantum Information Science and Engineering Network. https://qisenet.uchicago.edu/overview.

Quester, George H. *Nuclear First Strike: Consequences of a Broken Taboo.* Baltimore: Johns Hopkins University Press, 2006.

Rasmussen, Richard. *JTOT Health and Safety Considerations.* LA-UR-18-25517, Los Alamos National Laboratory. June 22, 2018. https://permalink.lanl.gov/object/tr?what=info:lanl-repo/lareport/LA-UR-18-25517.

Reed, Cameron. "The Nuclear Emergency Support Team (NEST)." *Physics and Society* 42, no. 1 (January 2013).

Reed, Mary Beth. *You Can't Run a Reactor if You Can't Get to It: A Study of Savannah River Site's Infrastructure.* Stone Mountain, GA: New South Associates, 2010.

Reeves, Geoffrey D., Josef Koller, Robert L. Tokar, Yue Chen, Michael G. Henderson, and Reiner H. Friedel. "The Dynamic Radiation Environment Assimilation Model (DREAM)." LA-UR-10-01851; LA-UR-10-1851, Los Alamos National Laboratory. Office of Scientific and Technical Information. January 1, 2010. https://www.osti.gov/biblio/1024361.

Reif, Kingston. "500,000,000,000 Reasons to Scrutinize the US Plan for Nuclear Weapons." *Bulletin of the Atomic Scientists.* May 21, 2021. https://thebulletin.org/2021/05/500000000000-reasons-to-scrutinize-the-us-plan-for-nuclear-weapons.

———. "Pentagon Raises Concerns About NNSA Budget." Arms Control Association. September 2021. https://www.armscontrol.org/act/2021-09/news/pentagon-raises-concerns-about-nnsa-budget.

Reif, Kingston, and Shannon Bugos. "President Biden's First Nuclear Arms Budget Is Disappointing." IDN—InDepthNews. July 17, 2021. https://www.indepthnews.net/index.php/the-world/usa-and-canada/4579-president-biden-s-first-nuclear-arms-budget-is-disappointing.

Richelson, Jeffrey. *Defusing Armageddon: Inside NEST, America's Secret Nuclear Bomb Squad.* New York: W. W. Norton, 2009.

Roberts, Brad. *The Case for U.S. Nuclear Weapons in the 21st Century.* 1st ed. Stanford, CA: Stanford University Press, 2015.

———."Stockpile Stewardship in an Era of Renewed Strategic Competition." Center for Global Security Research. 2022. https://cgsr.llnl.gov/content/assets/docs /CGSR_Occasional_Stockpile-Stewardship-Era-Renewed-Competition.pdf.

Rofer, Cheryl. "Plutonium Pits Are a Critical Obstacle in U.S. Nuclear Plans." *Foreign Policy.* August 9, 2021. https://foreignpolicy.com/2021/08/09/plutonium -pits-nuclear-plans-policy.

Romero, Christopher. *Final Design Review Report, SCE GEN 2, 3-Ft Confinement Vessel Weldment: Appendix A.* Office of Scientific and Technical Information. September 26, 2017. https://www.osti.gov/biblio/1396104-final-design-review -report-subcritical-experiments-gen-ft-confinement-vessel-weldment.

Roth, Nickolas, Matthew Bunn, and William H. Tobey. "Public Testimony for the Record FY 2021 House Committee on Appropriations Subcommittee on Energy and Water Development and Related Agencies FY 2021 Public Witness Hearing." Arms Control Center. March 31, 2020. https://armscontrolcenter.org/wp-content /uploads/2020/04/Roth-Tobey-Bunn-Testimony11.pdf.

Sandia National Laboratories. "Rare Open-Access Quantum Computer Now Operational." Sandia National Laboratories. March 15, 2021. https://newsreleases.sandia .gov/quantum_testbed.

———. "Sandia Virtual Tours." Sandia National Laboratories. https://tours.sandia.gov /tours.html.

———. "70 Ways Sandia Has Changed the Nation." Sandia National Laboratories. https://www.sandia.gov/about/70-ways.

Sauder, Richard. *Underground Bases and Tunnels: What Is the Government Trying to Hide?* Kempton, IL: Adventures Unlimited Press, 2015.

Savannah River Site. "Plutonium Pit Production at SRS." 2022. Savannah River Site. https://www.srs.gov/general/news/factsheets/SRS_SRPPF_2022.pdf.

Scarlett, Harry. "Nuclear Weapon Culture." LA-UR-20-2505, Los Alamos National Laboratory. https://permalink.lanl.gov/object/tr?what=info:lanl-repo/lareport /LA-UR-20-25059.

Schell, Jonathan. *The Seventh Decade: The New Shape of Nuclear Danger.* 1st ed. New York: Metropolitan Books/Henry Holt and Co., 2007.

Schlosser, Eric. *Command and Control: Nuclear Weapons, the Damascus Accident, and the Illusion of Safety.* New York: Penguin Books, 2014.

Schweber, Silvan S. *In the Shadow of the Bomb: Oppenheimer, Bethe, and the Moral Responsibility of the Scientist.* Princeton, NJ: Princeton University Press, 2007.

Scooby. *LLNL: The True Story.* http://llnlthetruestory.blogspot.com.

Shah, Agam. "Cerebras Chip Part of Project to Spot Post-exascale Technology." *HPCWire.* October 9, 2022. https://www.hpcwire.com/2022/10/19/cerebras -chip-part-of-project-to-spot-post-exascale-technology.

Shultz, George P., William J. Perry, Henry A. Kissinger, and Sam Nunn. "A World Free of Nuclear Weapons." *Wall Street Journal.* January 4, 2007. https://www.wsj.com /articles/SB116787515251566636.

Shultz, George Pratt. *Nuclear Security: The Problems and the Road Ahead*. Stanford, CA: Hoover Institution Press, 2014.

sirapwm. "Singularity in Rutherford Cross Section." Physics Forums. August 23, 2021. https://www.physicsforums.com/threads/singularity-in-rutherford-cross-section.371737.

Sirota, Sara. "Congress Is Already Blowing a Key Chance to Reform Nuclear Weapons Policy." *The Intercept*. March 24, 2022. https://theintercept.com/2022/03/24/nuclear-weapons-reform-commission-ukraine-russia.

Smith, Jeffrey R. "Sensors Add to Accuracy and Power of U.S. Nuclear Weapons but May Create New Security Perils." *Washington Post*. October 29, 2021. www.washingtonpost.com. https://www.washingtonpost.com/national-security/us-nuclear-weapons-electronic-sensors-accuracy/2021/10/28/79533ff0-34cc-11ec-9bc4-86107e7b0ab1_story.html.

Smith, Peter D. *Doomsday Men: The Real Dr. Strangelove and the Dream of the Superweapon*. New York: St. Martin's Press, 2007.

Snelson, Catherine M., Christopher R. Bradley, William R. Walter, Tarabay H. Antoun, Robert Abbott, Kyle Jones, Veraun D. Chipman, et al. "The Source Physics Experiment (SPE) Science Plan (Version 2.4c)." LLNL-TR-654513, Lawrence Livermore National Laboratory. Office of Scientific and Technical Information. June 27, 2019. doi:10.2172/1529820.

Sterngold, James. "Los Alamos Scientist Criticizes Federal Approach to Arsenal." *SFGATE*. February 13, 2007. https://www.sfgate.com/news/article/Los-Alamos-scientist-criticizes-federal-approach-2649254.php.

Straub, Mark D., John Arnold, Julianna Fessenden, and Jaqueline L. Kiplinger. "Recent Advances in Nuclear Forensic Chemistry." *Analytical Chemistry* 93, no. 1 (January 2021): 3–22. doi:10.1021/acs.analchem.0c03571.

Szasz, Ferenc Morton. *The Day the Sun Rose Twice: The Story of the Trinity Site Nuclear Explosion, July 16, 1945*. 1st ed. Albuquerque: University of New Mexico Press, 1984.

Taylor, Bryan C., and Judith Hendry. "Insisting on Persisting: The Nuclear Rhetoric of 'Stockpile Stewardship.'" *Rhetoric & Public Affairs* 11, no. 2 (2008): 303–334. doi:10.1353/rap.0.0040.

Terry, Russell. *Opportunities in Space Science: An Overview of Hard Radiation Sensing Applications and Research in ISR-1* . LA-UR-12-01271, Los Alamos National Laboratory. March 13, 2012. https://permalink.lanl.gov/object/tr?what=info:lanl-repo/lareport/LA-UR-12-01271.

"36 Cerro Grande Fire Premium High Res Photos." Getty Images. https://www.gettyimages.com/photos/cerro-grande-fire (accessed November 2, 2022).

Tilden, Jay. "DOE/NNSA Nuclear Emergency Support Team Offsite Activities." US Nuclear Regulatory Commission. May 5, 2020. https://www.nrc.gov/docs/ML2013/ML20136A015.pdf.

Tollefson, Jeff. "US Achieves Laser-Fusion Record: What It Means for Nuclear-Weapons Research." *Nature* 597, no. 7875 (August 2021): 163–164. doi:10.1038/d41586-021-02338-4.

"U1A Complex." Nevada National Security Site. http://www.nnss.gov/pages/facilities/U1aComplex.html.

"Underground Bases and Tunnels." *The "New World Order."* August 2016. https://thenwwrdor.blogspot.com/2016/08/photo-of-united-states-air-force-tunnel.html.

US Department of Energy, Office of Inspector General Office of Audits and Inspection. "Audit Report: OAS-L-14-09: National Nuclear Security Administration's Space-Based Nuclear Detonation Detection Program." OAS-L-14-0, Department of Energy. July 2014. https://www.energy.gov/ig/articles/audit-report-oas-1 -14-09.

US Fish and Wildlife Service. "Rocky Flats." US Fish and Wildlife Service. https:// www.fws.gov/refuge/rocky_flats.

US Government Printing Office. *The United States Strategic Bombing Survey: The Effects of Atomic Bombs on Hiroshima and Nagasaki, June 30, 1946*. RWU Archives and Special Collections, 1946.

US Nuclear Deterrence Policy and Strategy 2. June 16, 2021. https://web cache.googleusercontent.com/search?q=cache:w_evMjkT-HkJ:https://www .armed-services.senate.gov/hearings/united-states-nuclear-deterrence-policy-and -strategy-2+&cd=2&hl=en&ct=clnk&gl=us.

US Senate Committee on Appropriations. "A Review of the Fiscal Year 2023 Budget Submission for National Nuclear Security Administration." US Senate Committee on Appropriations. May 18, 2022. https://www.appropriations.senate.gov /hearings/a-review-of-the-fiscal-year-2023-budget-submission-for-national-nuclear -security-administration.

Vickery, Jonathan. "SRS Lawsuit Ends in $600 Million Settlement." *Augusta Chronicle*. September 3, 2020. https://www.augustachronicle.com/story/news/2020/09/03 /srs-lawsuit-ends-in-600-million-settlement/114788298.

Weart, Spencer R. *Nuclear Fear: A History of Images*. Cambridge, MA: Harvard University Press, 1988.

White House. "National Security Strategy of the United States of America." White House. December 2017. https://www.whitehouse.gov/wp-content/uploads/2017/12 /NSS-Final-12-18-2017-0905.pdf.

Win Without War. "Principles of a Progressive Foreign Policy for the United States." Win Without War. https://winwithoutwar.org/policy/principles-of-a-progressive -foreign-policy-for-the-united-states (accessed August 18, 2021).

Winsberg, Eric. "Models of Success Versus the Success of Models: Reliability Without Truth." *Synthese* 152, no. 1 (September 2006): 1–19. doi:10.1007/s11229-004 -5404-6.

———. "Sanctioning Models: The Epistemology of Simulation." *Science in Context* 12, no. 2 (1999): 275–292. doi:10.1017/S0269889700003422.

———. *Science in the Age of Computer Simulation*. Chicago: University of Chicago Press, 2010.

———. "Simulated Experiments: Methodology for a Virtual World." *Philosophy of Science* 70, no. 1 (2003): 105–125. doi:10.1086/367872.

Wolfe, Audra J. *Competing with the Soviets: Science, Technology, and the State in Cold War America*. Baltimore: Johns Hopkins University Press, 2013.

Wolfsthal, John, Kingston Reif, Richard Nephew, Kelsey Davenport, Alexandra Bell, Nickolas Roth, and Sharon Squassoni. "Blundering Toward Nuclear Chaos." Belfer Center for Science and International Affairs. May 2020. https://www.belfercenter .org/publication/blundering-toward-nuclear-chaos.

Wyland, Scott. "Biden Proposes $1 Billion for Nuclear Weapons Work." *Santa Fe New Mexican/Yahoo News.* May 29, 2021. https://news.yahoo.com/biden-proposes -1-billion-nuclear-233300597.html.

———. "Energy Department to Spend $15.5M to Upgrade Route from Los Alamos Lab to Waste Site." *Santa Fe New Mexican.* December 4, 2021. https://www .santafenewmexican.com/news/local_news/energy-department-to-spend-15-5m -to-upgrade-route-from-los-alamos-lab-to-waste/article_b8641ea2-53a9-11ec-9dc0 -7b311a21828f.html.

———. "LANL Workers Contaminated in Radiation Leak." *Santa Fe New Mexican.* January 28, 2022. https://www.santafenewmexican.com/news/local_news /lanl-workers-contaminated-in-radiation-leak/article_227ee5b4-8097-11ec-8452 -577c936c156d.html.

———. "LANL Would Get over $1B Bump in Proposed Budget." *Santa Fe New Mexican.* April 20, 2022. https://news.yahoo.com/lanl-over-1b-bump-proposed -150300706.html.

———. "LANL's Pit Production a Year Behind Schedule." *Santa Fe New Mexican.* October 4, 2022. https://www.santafenewmexican.com/news/local_news /lanls-pit-production-a-year-behind-schedule/article_ca2262a2-43f9-11ed-84da -eb276b365663.html.

———. "More Radioactive Contaminants Found at Los Alamos Housing Site." *Santa Fe New Mexican.* June 17, 2020. https://www.santafenewmexican.com/news /local_news/more-radioactive-contaminants-found-at-los-alamos-housing-site /article_06ea2152-aff0-11ea-8029-1b3f47747af9.html.

———. "Safety Questions Arise as Los Alamos National Laboratory Pursues Pit Production." *Santa Fe New Mexican.* May 6, 2022. https://www.santafenewmexican .com/news/local_news/safety-questions-arise-as-los-alamos-national-laboratory -pursues-pit-production/article_0fb60822-c651-11ec-9d78-8f2a8e8b2a10.html.

———. "Two More Radioactive Releases Reported at LANL." *Santa Fe New Mexican.* May 8, 2022. https://www.santafenewmexican.com/news/local_news /two-more-radioactive-releases-reported-at-lanl/article_2dd5d9fc-9e35-11ec-a88b -3ff0a5b61759.html.

## Interviews

Suzanne Ali, video interview to Sarah Scoles, January 2022.

Aaron Arnold, phone interview to Sarah Scoles, June 2021.

Andrew Baczewski, interview at Sandia National Laboratories to Sarah Scoles, October 2021.

Benjamin Bahney, video interview to Sarah Scoles, November 2021.

Laura Berzak Hopkins, interview at Lawrence Livermore National Laboratory to Sarah Scoles, May 2022.

Ed Bielejec, interview at Sandia National Laboratories to Sarah Scoles, October 2021.

Jason Bolles, interview at Sandia National Laboratories to Sarah Scoles, October 2021.

Bob Bonnett, interview at the Savannah River Site to Sarah Scoles, June 2022.

Jay Brotz, interview at Sandia National Laboratories to Sarah Scoles, October 2021.

Matthew Burger, interview at Sandia National Laboratories to Sarah Scoles, October 2021.

Josh Carmichael, interviews at Los Alamos National Laboratory to Sarah Scoles, August 2021, March 2022, and September 2022.

David Chavez, interview at Los Alamos National Laboratory to Sarah Scoles, August 2021.

Emma Claire Foley, phone interview to Sarah Scoles, September 2022.

David Clark, interview at Los Alamos National Laboratory and video interview to Sarah Scoles, August 2021 and May 2022.

Susan Clark, interview at Sandia National Laboratories to Sarah Scoles, October 2021.

Tom Clements, phone interview to Sarah Scoles, June 2021.

Jay Coghlan, phone interview to Sarah Scoles, July 2021.

Paul Crozier, interview at Sandia National Laboratories to Sarah Scoles, October 2021.

Nathan DeBardeleben, interview at Los Alamos National Laboratory to Sarah Scoles, August 2021.

Joanie Dix, email interview to Sarah Scoles, September 2021.

Arden Dougan, email interview to Sarah Scoles, September 2021.

Chris Fryer, interviews at Los Alamos National Laboratory to Sarah Scoles, August 2021, March 2022, and September 2022.

Arjun Gambhir, video interview to Sarah Scoles, November 2021.

Lee Glascoe, interview at Lawrence Livermore National Laboratory to Sarah Scoles, May 2022.

Bethany Goldblum, video interviews to Sarah Scoles, June 2021 and July 2022.

Rita Gonzalez, interview at Sandia National Laboratories to Sarah Scoles, October 2021.

Hugh Gusterson, phone interview to Sarah Scoles, May 2021.

Don Hancock, phone interview to Sarah Scoles, June 2021.

Anne I. Harrington, video interview to Sarah Scoles, July 2021.

Dan Haylett, video interview to Sarah Scoles, December 2021.

Jake Hecla, phone interview to Sarah Scoles, July 2021.

James Henkel, email interview to Sarah Scoles, September 2021.

Tina Hernandez, interview at Sandia National Laboratories to Sarah Scoles, October 2021.

John Hewson, interview at Sandia National Laboratories to Sarah Scoles, October 2021.

Vincent Ialenti, video interview to Sarah Scoles, May 2021.

Crystal Jaing, video interview to Sarah Scoles, November 2022.

Kevin Jameson, interview at Sandia National Laboratories to Sarah Scoles, October 2021.

Richard Jepsen, interview at Sandia National Laboratories to Sarah Scoles, October 2021.

Deana Kahnke, interview at Lawrence Livermore National Laboratory to Sarah Scoles, May 2022.

Virginia Kay, email interview to Sarah Scoles, September 2021.

Marylia Kelley, phone interview to Sarah Scoles, August 2021.

Ruth Kips, interview at Lawrence Livermore National Laboratory to Sarah Scoles, May 2022.

Scott Klenke, interview at Sandia National Laboratories to Sarah Scoles, October 2021.

Kalie Knecht, video interview to Sarah Scoles, June 2021.

Matt Korda, video interview to Sarah Scoles, July 2021.

Christopher Landers, email interview to Sarah Scoles, September 2021.

Alex Larzelere, video interview to Sarah Scoles, May 2021.

Tess Light, interviews at Los Alamos National Laboratory to Sarah Scoles, August 2021, March 2022, and September 2022.

Carlos Lopez, interview at Sandia National Laboratories to Sarah Scoles, October 2021.

Naomi Marks, interview at Lawrence Livermore National Laboratory to Sarah Scoles, May 2022.

Joseph Martz, interview at Los Alamos National Laboratory to Sarah Scoles, August 2021.

Eric Matthews, phone interview to Sarah Scoles, June 2021.

Luther McDonald, phone interview to Sarah Scoles, June 2021.

Laura McGill, interview at Sandia National Laboratories to Sarah Scoles, October 2021.

Nathan Michael, interview at Sandia National Laboratories to Sarah Scoles, October 2021.

Amir Mohaghegi, interview at Sandia National Laboratories to Sarah Scoles, October 2021.

Stephen Myers, interview at Lawrence Livermore National Laboratory to Sarah Scoles, May 2022.

Rob Neely, interview at Lawrence Livermore National Laboratory to Sarah Scoles, May 2022.

Mark Paris, interviews at Los Alamos National Laboratory to Sarah Scoles, August 2021, March 2022, and September 2022.

Martin Pfeiffer, video interview to Sarah Scoles, July 2022.

Kendall Pierson, interview at Sandia National Laboratories to Sarah Scoles, October 2021.

Miles Pomper, phone interview to Sarah Scoles, June 2021.

Terri Poxon-Pearson, email interview to Sarah Scoles, September 2021.

Terri Quinn, interview at Lawrence Livermore National Laboratory to Sarah Scoles, May 2022.

Jessica Rahim, email interview to Sarah Scoles, September 2021.

Brad Roberts, interview at Lawrence Livermore National Laboratory to Sarah Scoles, May 2022.

Ed Romero, interview at Sandia National Laboratories to Sarah Scoles, October 2021.

Nickolas Roth, phone interview to Sarah Scoles, June 2021.

Phil Schneider, video interview to Sarah Scoles, September 2022.

Jen Shafer, video interview to Sarah Scoles, July 2021.

Dan Sinars, interview at Sandia National Laboratories to Sarah Scoles, October 2021.

June Stanley, interview at Sandia National Laboratories to Sarah Scoles, October 2021.

Mark Straub, phone interview to Sarah Scoles, July 2021.

Michael Thompson, email interview to Sarah Scoles, September 2021.

Kevin Veal, email interview to Sarah Scoles, September 2021.

Bill Wanderer, email interview to Sarah Scoles, October 2021.

Sharon Weiner, video interview to Sarah Scoles, July 2021.

Alex Wellerstein, phone interview to Sarah Scoles, July 2021.

Alicia Williams, interview at Lawrence Livermore National Laboratory to Sarah Scoles, May 2022.

Geoff Wilson, phone interview to Sarah Scoles, June 2021.

Eric Winsberg, phone interview to Sarah Scoles, May 2021.

Walt Witkowski, interview at Sandia National Laboratories to Sarah Scoles, October 2021.

Audra Wolfe, video interview to Sarah Scoles, August 2021.

# INDEX